Complex Variables Demystified

Demystified Series

Accounting Demystified
Advanced Calculus Demystified
Advanced Physics Demystified
Advanced Statistics Demystified
Algebra Demystified
Alternative Energy Demystified
Anatomy Demystified
Astronomy Demystified
Audio Demystified
Biochemistry Demystified
Biology Demystified
Biotechnology Demystified
Business Calculus Demystified
Business Math Demystified
Business Statistics Demystified
C++ Demystified
Calculus Demystified
Chemistry Demystified
Circuit Analysis Demystified
College Algebra Demystified
Complex Variables Demystified
Corporate Finance Demystified
Databases Demystified
Diabetes Demystified
Differential Equations Demystified
Digital Electronics Demystified
Earth Science Demystified
Electricity Demystified
Electronics Demystified
Engineering Statistics Demystified
Environmental Science Demystified
Everyday Math Demystified
Fertility Demystified
Financial Planning Demystified
Forensics Demystified
French Demystified
Genetics Demystified
Geometry Demystified
German Demystified
Global Warming and Climate Change Demystified
Hedge Funds Demystified
Investing Demystified
Italian Demystified
Java Demystified
JavaScript Demystified
Lean Six Sigma Demystified
Linear Algebra Demystified
Macroeconomics Demystified
Management Accounting Demystified
Math Proofs Demystified
Math Word Problems Demystified
MATLAB® Demystified
Medical Billing and Coding Demystified
Medical-Surgical Nursing Demystified
Medical Terminology Demystified
Meteorology Demystified
Microbiology Demystified
Microeconomics Demystified
Nanotechnology Demystified
Nurse Management Demystified
OOP Demystified
Options Demystified
Organic Chemistry Demystified
Pharmacology Demystified
Physics Demystified
Physiology Demystified
Pre-Algebra Demystified
Precalculus Demystified
Probability Demystified
Project Management Demystified
Psychology Demystified
Quantum Field Theory Demystified
Quantum Mechanics Demystified
Real Estate Math Demystified
Relativity Demystified
Robotics Demystified
Sales Management Demystified
Signals and Systems Demystified
Six Sigma Demystified
Spanish Demystified
SQL Demystified
Statics and Dynamics Demystified
Statistics Demystified
Technical Analysis Demystified
Technical Math Demystified
Trigonometry Demystified
Vitamins and Minerals Demystified

Complex Variables Demystified

David McMahon

New York Chicago San Francisco Lisbon London
Madrid Mexico City Milan New Delhi San Juan
Seoul Singapore Sydney Toronto

The McGraw-Hill Companies

Library of Congress Cataloging-in-Publication Data

McMahon, David.
Complex variables demystified / David McMahon.
p. cm.
ISBN 978-0-07-154920-2 (alk. paper)
1. Functions of complex variables. I. Title.
QA331.7.M42 2008
515′.9—dc22 2008023102

1 2 3 4 5 6 7 8 9 0 FGR/FGR 0 1 4 3 2 1 0 9 8

ISBN 978-0-07-154920-2
MHID 0-07-154920-X

Sponsoring Editor
Judy Bass

Production Supervisor
Pamela A. Pelton

Editing Supervisor
Stephen M. Smith

Project Manager
Harleen Chopra, International Typesetting and Composition

Copy Editor
Ragini Pandey, International Typesetting and Composition

Proofreader
Bhavna Gupta, International Typesetting and Composition

Indexer
WordCo Indexing Services, Inc.

Art Director, Cover
Jeff Weeks

Composition
International Typesetting and Composition

Printed and bound by Quebecor/Fairfield.

ABOUT THE AUTHOR

David McMahon has worked for several years as a physicist and researcher at Sandia National Laboratories. He is the author of *Linear Algebra Demystified, Quantum Mechanics Demystified, Relativity Demystified, MATLAB® Demystified,* and *Quantum Field Theory Demystified*, among other successful titles.

CONTENTS

PREFACE

Complex variables, and its more advanced version, *complex analysis,* is one of the most fascinating areas in pure and applied mathematics. It all started when mathematicians were mystified by equations that could only be solved if you could take the square roots of negative numbers. This seemed bizarre, and back then nobody could imagine that something as strange as this could have any application in the real world. Thus the term *imaginary number* was born and the area seemed so odd it became known as *complex.*

These terms have stuck around even though the theory of complex variables has found a home as a fundamental part of mathematics and has a wide range of physical applications. In mathematics, it turns out that complex variables are actually an extension of the real variables.

A student planning on becoming a professional pure or applied mathematician should definitely have a thorough grasp of complex analysis.

Perhaps the most surprising thing about complex variables is the wide range of applications it touches in physics and engineering. In many of these applications, complex variables proves to be a useful tool. For example, because of *Euler's identity,* a formula we use over and over again in this book, electromagnetic fields are easier to deal with using complex variables.

Other areas where complex variables plays a role include fluid dynamics, the study of temperature, electrostatics, and in the evaluation of many real integrals of functions of a real variable.

In quantum theory, we meet the most surprising revelation about complex variables. It turns out they are not so imaginary at all. Instead, they appear to be as "real" as real numbers and even play a fundamental role in the working of physical systems at the microscopic level.

In the limited space of this book, we won't be able to cover the physical applications of complex variables. Our purpose here is to build a solid foundation to get you started on the subject. This book is filled with a large number of solved examples (many of which are at the advanced undergraduate level) that will show you how to tackle problems in complex variables, with explicit detail.

Topics covered include:

- Complex numbers, variables, and the polar representation
- Limits and continuity
- Derivatives and the Cauchy-Riemann equations
- Elementary functions like the exponential and trigonometric functions
- Complex integration
- The residue theorem
- Conformal mapping
- Sequences, infinite series, and Laurent series
- The gamma and zeta functions
- Solving boundary value problems

This book should provide the reader with a good introduction to the subject of complex variables. After completing this book, you will be able to deepen your knowledge of the subject by consulting one of the excellent texts listed in the references at the end of the book.

I would like to thank Steven G. Krantz for his very thoughtful and thorough review of this manuscript.

David McMahon

CHAPTER 1

Complex Numbers

In the early days of modern mathematics, people were puzzled by equations like this one:

$$x^2 + 1 = 0$$

The equation looks simple enough, but in the sixteenth century people had no idea how to solve it. This is because to the common-sense mind the solution seems to be without meaning:

$$x = \pm\sqrt{-1}$$

For this reason, mathematicians dubbed $\sqrt{-1}$ an *imaginary number*. We abbreviate this by writing "i" in its place, that is:

$$i = \sqrt{-1} \tag{1.1}$$

So we see that $i^2 = -1$, and we can solve equations like $x^2 + 1 = 0$. Note that electrical engineers use $j = \sqrt{-1}$, but we will stick with the standard notation used in mathematics and physics.

The Algebra of Complex Numbers

More general complex numbers can be written down. In fact, using real numbers a and b we can form a complex number:

$$c = a + ib \tag{1.2}$$

We call a the *real part* of the complex number c and refer to b as the *imaginary part* of c. The numbers a and b are ordinary real numbers. Now let $c = a + ib$ and $k = m + in$ be two complex numbers. Here m and n are also two arbitrary real numbers (not integers, we use m and n because I am running out of symbols to use). We can form the sum and difference of two complex numbers by adding (subtracting) their real and imaginary parts independently. That is:

$$c + k = a + ib + m + in = (a + m) + i(b + n)$$
$$c - k = a + ib - (m + in) = (a - m) + i(b - n)$$

To multiply two complex numbers, we just multiply out the real and imaginary parts term by term and use $i^2 = -1$, then group real and imaginary parts at the end:

$$\begin{aligned} ck &= (a + ib)(m + in) = am + ian + ibm + i^2 bn \\ &= am + ian + ibm - bn \\ &= (am - bn) + i(an + bm) \end{aligned}$$

To divide two complex numbers and write the result in the form $c = a + ib$, we're going to need a new concept, called the *complex conjugate.* We form the complex conjugate of any complex number by letting $i \to -i$. The complex conjugate is indicated by putting a bar on top of the number or variable. Again, let $c = a + ib$. Then the complex conjugate is

$$\bar{c} = a - ib \tag{1.3}$$

It's easy to see that if c is purely real, that is, $c = a$, then the complex conjugate is $\bar{c} = \bar{a} = a$. On the other hand, if c is purely imaginary, then $c = ib$. This means that $\bar{c} = \overline{ib} = -ib = -c$. Taking the complex conjugate twice gives back the original number:

$$\bar{\bar{c}} = \overline{a - ib} = a + ib = c$$

Notice what happens when we multiply a complex number by its conjugate:

$$c\bar{c} = (a+ib)(a-ib) = a^2 - iab + iba - i^2b^2$$
$$= a^2 - i^2b^2 = a^2 + b^2$$

We call the quantity $c\bar{c}$ the *modulus* of the complex number c and write

$$|c|^2 = c\bar{c} \tag{1.4}$$

Note that in physics, the complex conjugate is often denoted by an asterisk, that is, c^*. The modulus of a complex number has geometrical significance. This is because we can view a complex number as a vector in the plane with components given by the real and imaginary parts. The length of the vector corresponds to the modulus. We will discuss this concept again later (see Fig. 1.1).

Now we can find the result of c/k, provided that $k \neq 0$ of course. We have

$$\frac{c}{k} = \frac{a+ib}{m+in}$$
$$= \frac{a+ib}{m+in}\frac{(m-in)}{(m-in)}$$
$$= \frac{am+ibm-ian+bn}{m^2+n^2}$$
$$= \frac{am+bn}{m^2+n^2} + i\frac{bm-an}{m^2+n^2}$$

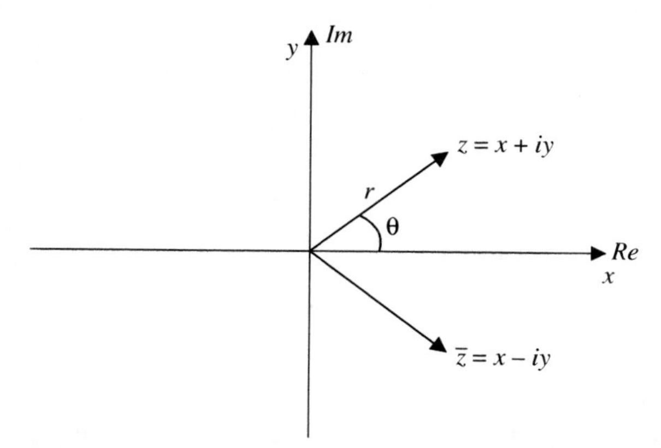

Figure 1.1 The complex plane, showing $z = x + iy$ and its complex conjugate as vectors.

We say that two complex numbers are equal if and only if their real and imaginary parts are equal. That is, $c = a + ib$ and $k = m + in$ are equal if and only if

$$a = m \qquad b = n$$
$$\Rightarrow c = k$$

Complex Variables

In the early days, all of this probably seemed like a neat little trick that could be used to solve obscure equations, and not much more than that. But in reality it opened up a Pandora's box of possibilities that is still being dealt with today. It turns out that an entire branch of analysis called *complex analysis* can be constructed, which really supersedes real analysis. Complex analysis has not only transformed the world of mathematics, but surprisingly, we find its application in many areas of *physics and engineering.* For example, we can use complex numbers to describe the behavior of the electromagnetic field. In atomic systems, which are described by quantum mechanics, complex numbers and *complex functions* play a central role, and actually appear to be a fundamental part of nature. Complex numbers are often hidden. For example, as we'll see later, the trigonometric functions can be written down in surprising ways like:

$$\cos\theta = \frac{e^{i\theta} + e^{-i\theta}}{2} \qquad \sin\theta = \frac{e^{i\theta} + e^{-i\theta}}{2i}$$

It appears that complex numbers are not so "imaginary" after all; rather they are used in a wide variety of engineering and science applications.

The first step in moving forward toward a calculus based on complex numbers is to abstract the notion of a complex number to a *complex variable.* This is the same as abstracting the notion of a real number to a variable like x that we can use to solve algebraic equations. We use z to represent a *complex variable.* Its real and imaginary parts are represented by the real variables x and y, respectively. So we write

$$z = x + iy \tag{1.5}$$

The complex conjugate is then

$$\bar{z} = x - iy$$

A complex number and its conjugate have an interesting origin in the study of polynomials with real coefficients. Let p be a polynomial with real coefficients and suppose that a complex number z is a root of p. Then it follows that the complex conjugate $\bar{z}$ is a root of p also.

The modulus of the complex variable z is given by

$$|z|^2 = x^2 + y^2 \qquad \Rightarrow |z| = \sqrt{x^2 + y^2} \tag{1.6}$$

The same rules for addition, subtraction, multiplication, and division we illustrated with complex numbers apply to complex variables. So if $z = x + iy$ and $w = u + iv$ then

$$zw = (x + iy)(u + iv) = (xu - yv) + i(yu + xv)$$

We can graph complex numbers in the x-y plane, which we sometimes call the *complex plane* or the *z plane*. The y axis is the *imaginary axis* and the x axis is the *real axis*. A complex number $z = x + iy$ can be depicted as a vector in the complex plane, with a length r given by its modulus:

$$r = |z| = \sqrt{x^2 + y^2} \tag{1.7}$$

We also keep track of the angle θ that this vector makes with the real axis. The complex conjugate is a vector reflected across the real axis. This is easy to understand since we form the conjugate by letting $y \to -y$. These ideas are illustrated in Fig. 1.1.

Rules for the Complex Conjugate

Let $z = x + iy$ and $w = u + iv$ be two complex variables. Then

$$\begin{aligned} \overline{z + w} &= \bar{z} + \bar{w} \\ \overline{zw} &= \bar{z}\,\bar{w} \\ \overline{\left(\frac{z}{w}\right)} &= \frac{\bar{z}}{\bar{w}} \end{aligned} \tag{1.8}$$

These properties are easy to demonstrate. For example, we prove the first one:

$$\begin{aligned} \overline{z + w} &= \overline{(x + iy) + (u + iv)} \\ &= \overline{(x + u) + i(y + v)} \\ &= (x + u) - i(y + v) \\ &= x - iy + u - iv \\ &= \bar{z} + \bar{w} \end{aligned}$$

If $z \neq 0$, we can form the multiplicative inverse of z which we denote by z^{-1}. The inverse has the property that

$$zz^{-1} = 1 \tag{1.9}$$

It is given by

$$z^{-1} = \frac{\bar{z}}{|z|^2} = \frac{1}{z} \tag{1.10}$$

We can verify that this works explicitly in two ways:

$$zz^{-1} = z\frac{\bar{z}}{|z|^2} = \frac{|z|^2}{|z|^2} = 1$$
$$zz^{-1} = (x+iy)\frac{x-iy}{x^2+y^2} = \frac{x^2+y^2}{x^2+y^2} = 1$$

Notice that the inverse gives us a way to write the quotient of two complex numbers, allowing us to do division:

$$\frac{z}{w} = \frac{z\bar{w}}{w\bar{w}} = \frac{z\bar{w}}{|w|^2}$$

EXAMPLE 1.1

Find the complex conjugate, sum, product, and quotient of the complex numbers

$$z = 2 - 3i \qquad w = 1 + i$$

SOLUTION

To find the complex conjugate of each complex number we let $i \to -i$. Hence

$$\bar{z} = \overline{2-3i} = 2 + i3$$
$$\bar{w} = \overline{1+i} = 1 - i$$

The sum of the two complex numbers is formed by adding the real and imaginary parts, respectively:

$$z + w = (2 - 3i) + (1 + i) = (2 + 1) + i(-3 + 1) = 3 - 2i$$

We can form the product as follows:

$$\begin{aligned} zw &= (2 - 3i)(1 + i) \\ &= 2 - 3i + 2i - 3i^2 \\ &= (2 + 3) + i(-3 + 2) \\ &= 5 - i \end{aligned}$$

Finally, we use the complex conjugate of w to form the quotient:

$$\frac{z}{w} = \frac{z}{w}\left(\frac{\overline{w}}{\overline{w}}\right) = \frac{(2 - 3i)}{(1 + i)}\frac{(1 - i)}{(1 - i)} = \frac{2 - 3i - 2i + 3i^2}{1 + i - i - i^2} = \frac{2 - 3 - 5i}{1 + 1} = \frac{-1 - 5i}{2}$$

EXAMPLE 1.2

Earlier, we said that if $z = x + iy$, then x is the real part of z [denoted by writing $x = \text{Re}(z)$ say] and that y is the imaginary part of z [$y = \text{Im}(z)$]. Derive expressions that allow us to define the real and imaginary parts of a complex number using only z and its complex conjugate.

SOLUTION

First let's write down the complex variable and its complex conjugate:

$$z = x + iy \qquad \overline{z} = x - iy$$

Now we see that this is just simple algebra. We can eliminate y from both equations by adding them:

$$z + \overline{z} = x + iy + x - iy = 2x$$

So, we find that the real part of z is given by

$$\text{Re}(z) = x = \frac{z + \overline{z}}{2} \tag{1.11}$$

Now, let's subtract the complex conjugate from z instead, which allows us to eliminate x:

$$z - \overline{z} = x + iy - (x - iy) = x + iy - x + iy = 2iy$$

$$\Rightarrow \text{Im}(z) = y = \frac{z - \overline{z}}{2i} \tag{1.12}$$

EXAMPLE 1.3

Find z^2 if $z = (2+i)/[4i - (1+2i)]$.

SOLUTION

Note that when the modulus sign is not present, we square without computing the complex conjugate. That is $|z|^2 = z\overline{z}$ but $z^2 = z \cdot z$, which is a different quantity. So in this case we have

$$\begin{aligned}
z^2 &= \left(\frac{2+i}{4i-(1+2i)}\right)^2 \\
&= \left(\frac{2+i}{4i-(1+2i)}\right)\left(\frac{2+i}{4i-(1+2i)}\right) \\
&= \left(\frac{4+4i+i^2}{(-1+2i)(-1+2i)}\right) \\
&= \left(\frac{3+4i}{-3-4i}\right) \\
&= \left[\frac{(3+4i)(-3+4i)}{(-3-4i)(-3+4i)}\right] \text{(multiply and divide by complex conjugate of denominator)} \\
&= \frac{-9-12i+12i-16}{9+12i-12i+16} = \frac{-25}{25} = -1
\end{aligned}$$

EXAMPLE 1.4

Show that $1/i = -i$.

SOLUTION

This is easy, using the rule we've been applying for division. That is:

$$\frac{z}{w} = \frac{z}{w}\left(\frac{\overline{w}}{\overline{w}}\right)$$

Hence

$$\frac{1}{i} = \frac{1}{i}\left(\frac{-i}{-i}\right) = \frac{-i}{-(i^2)} = \frac{-i}{-(-1)} = -i$$

EXAMPLE 1.5

Find z if $z(7z + 14 - 5i) = 0$.

SOLUTION

One obvious solution to the equation is $z = 0$. The other one is found to be

$$7z + 14 - 5i = 0$$
$$\Rightarrow 7z = -14 + 5i$$

or

$$z = -2 + \frac{5}{7}i$$

Pascal's Triangle

Expansions of complex numbers can be written down immediately using *Pascal's triangle,* which lists the coefficients in an expansion of the form $(x + y)^n$. We list the first five rows here:

$$\begin{array}{ccccccccc} & & & & 1 & & & & \\ & & & 1 & & 1 & & & \\ & & 1 & & 2 & & 1 & & \\ & 1 & & 3 & & 3 & & 1 & \\ 1 & & 4 & & 6 & & 4 & & 1 \end{array} \tag{1.13}$$

The first row corresponds to $(x + y)^0$, the second row to $(x + y)^1$, and so on. For example, looking at the third row we have coefficients 1, 2, 1. This means that

$$(x + y)^2 = x^2 + 2xy + y^2$$

EXAMPLE 1.6

Write $(2 - i)^4$ in the standard form $a + ib$.

SOLUTION

The coefficients for the fourth power are found in row five of Pascal's triangle. In general:

$$(x+y)^4 = x^4 + 4x^3y + 6x^2y^2 + 4xy^3 + y^4$$

Hence

$$\begin{aligned}(2-i)^4 &= 2^4 + 4(2^3)(-i) + 6(2^2)(-i)^2 + 4(2)(-i)^3 + (-i)^4 \\ &= 16 - 32i + 24(-i)^2 + 8(-i)^3 + (-i)^4\end{aligned}$$

Now let's look at some of the terms involving powers of $-i$ individually. First we have

$$(-i)^2 = (-1)^2 i^2 = (+1)(-1) = -1$$

The last two terms are

$$(-i)^3 = (-1)^3 i^3 = (-1)(i \cdot i^2) = (-1)(i)(-1) = +i$$

$$(-i)^4 = [(-i)^2]^2 = (-1)^2 = +1$$

Therefore we have

$$\begin{aligned}(2-i)^4 &= 16 - 32i + 24(-i)^2 + 8(-i)^3 + (-i)^4 \\ &= 16 - 32i - 24 + 8i + 1 \\ &= -7 - 24i\end{aligned}$$

Axioms Satisfied by the Complex Number System

We have already seen some of the basics of how to handle complex numbers, like how to add or multiply them. Now we state the formal axioms of the complex number system which allow mathematicians to describe complex numbers as a

field. These axioms should be familiar since their general statement is similar to that used for the reals. We suppose that u, w, z are three complex numbers, that is, $u, w, z \in \mathbb{C}$. Then these axioms follow:

$$z + w \quad \text{and} \quad zw \in \mathbb{C} \quad \text{(closure law)} \tag{1.14}$$

$$z + w = w + z \quad \text{(commutative law of addition)} \tag{1.15}$$

$$u + (w + z) = (u + w) + z \quad \text{(associative law of additio)} \tag{1.16}$$

$$zw = wz \quad \text{(commutative law of multiplication)} \tag{1.17}$$

$$u(wz) = (uw)z \quad \text{(associative law of multiplication)} \tag{1.18}$$

$$u(w + z) = uw + uz \quad \text{(distributive law)} \tag{1.19}$$

The identity with respect to addition is given by $z = 0 + 0i$, which satisfies

$$z + 0 = 0 + z \tag{1.20}$$

The identity with respect to multiplication is given by $z = 1 + i0 = 1$, which satisfies

$$z \cdot 1 = 1 \cdot z = z \tag{1.21}$$

For any complex number z there exists an *additive inverse,* which we denote by $-z$ that satisfies

$$z + (-z) = (-z) + z = 0 \tag{1.22}$$

There also exists a multiplicative inverse z^{-1}, which we have seen satisfies

$$zz^{-1} = z^{-1}z = 1 \tag{1.23}$$

A set that satisfies properties in Eqs. (1.14)–(1.23) is called a *field.* The algebraic closure property in Eq. (1.14) illustrates that you can add two complex numbers together and you get another complex number (that is what we mean by *closed*). The complex numbers are the smallest algebraically closed field that contains the reals as a subset.

Properties of the Modulus

We have already seen that the modulus or *magnitude* or *absolute value* of a complex number is defined by multiplying it by its complex conjugate and taking the positive square root. The absolute value operator satisfies several properties. Let $z_1, z_2, z_3, \ldots, z_n$ be complex numbers. Then

$$|z_1 z_2| = |z_1||z_2| \tag{1.24}$$

$$|z_1 z_2 z_3 \ldots z_n| = |z_1||z_2||z_3|\ldots|z_n| \tag{1.25}$$

$$\left|\frac{z_1}{z_2}\right| = \frac{|z_1|}{|z_2|} \tag{1.26}$$

A relationship called the *triangle inequality* deserves special attention:

$$|z_1 + z_2| \le |z_1| + |z_2| \tag{1.27}$$

$$|z_1 + z_2 + \cdots + z_n| \le |z_1| + |z_2| + \cdots + |z_n| \tag{1.28}$$

$$|z_1 + z_2| \ge |z_1| - |z_2| \tag{1.29}$$

$$|z_1 - z_2| \ge |z_1| - |z_2| \tag{1.30}$$

Also note that $w\bar{z} + \bar{z}w = 2\,\mathrm{Re}(z\bar{w}) \le 2|z||w|$.

The Polar Representation

In Fig. 1.1, we showed how a complex number can be represented by a vector in the x-y plane. Using polar coordinates, we can develop an equivalent *polar representation* of a complex number. We say that $z = x + iy$ is the *Cartesian representation* of a complex number. To write down the polar representation, we begin with the definition of the polar coordinates (r, θ):

$$x = r\cos\theta \qquad y = r\sin\theta \tag{1.31}$$

We have already seen that when we represent a complex number as a vector in the plane the length of that vector is r. Hence, carrying forward with the vector analogy, the modulus of z is given by

$$r = \sqrt{x^2 + y^2} = |x + iy| \tag{1.32}$$

Using Eq. (1.31), we can write $z = x + iy$ as

$$\begin{aligned} z = x + iy &= r\cos\theta + ir\sin\theta \\ &= r(\cos\theta + i\sin\theta) \end{aligned} \tag{1.33}$$

Note that $r > 0$ and that we have $\tan\theta = y/x$ as a means to convert between polar and Cartesian representations.

THE ARGUMENT OF Z

The value of θ for a given complex number is called the *argument* of z or *arg z*. The *principal value* of arg z which is denoted by Arg z is the value $-\pi < \Theta \leq \pi$. The following relationship holds:

$$\arg z = \text{Arg } z + 2n\pi \qquad n = 0, \pm 1, \pm 2, \ldots \tag{1.34}$$

The principal value can be specified to be between 0 and 2π.

EULER'S FORMULA

Euler's formula allows us to write the expression $\cos\theta + i\sin\theta$ in terms of a complex exponential. This is easy to see using a Taylor series expansion. First let's write out a few terms in the well-known Taylor expansions of the trigonometric functions cos and sin:

$$\cos\theta = 1 - \frac{1}{2}\theta^2 + \frac{1}{4!}\theta^4 - \frac{1}{6!}\theta^6 + \cdots \tag{1.35}$$

$$\sin\theta = \theta - \frac{1}{3!}\theta^3 + \frac{1}{5!}\theta^5 - \cdots \tag{1.36}$$

Now, let's look at $e^{i\theta}$. The power series expansion of this function is given by

$$
\begin{aligned}
e^{i\theta} &= 1 + i\theta + \frac{1}{2}(i\theta)^2 + \frac{1}{3!}(i\theta)^3 + \frac{1}{4!}(i\theta)^4 + \frac{1}{5!}(i\theta)^5 + \cdots \\
&= 1 + i\theta - \frac{1}{2}\theta^2 - i\frac{1}{3!}\theta^3 + \frac{1}{4!}\theta^4 + i\frac{1}{5!}\theta^5 + \cdots \\
&\quad \text{(Now group terms—looking for sin and cosine)} \\
&= \left(1 - \frac{1}{2}\theta^2 + \frac{1}{4!}\theta^4 - \cdots\right) + \left(i\theta - i\frac{1}{3!}\theta^3 + i\frac{1}{5!}\theta^5 + \cdots\right) \\
&= \left(1 - \frac{1}{2}\theta^2 + \frac{1}{4!}\theta^4 - \cdots\right) + i\left(\theta - \frac{1}{3!}\theta^3 + \frac{1}{5!}\theta^5 + \cdots\right) \\
&= \cos\theta + i\sin\theta
\end{aligned}
$$

So, we conclude that

$$e^{i\theta} = \cos\theta + i\sin\theta \tag{1.37}$$

$$e^{-i\theta} = \cos\theta - i\sin\theta \tag{1.38}$$

As noted in the introduction, these formulas can be inverted using algebra to obtain the following relationships:

$$\cos\theta = \frac{e^{i\theta} + e^{-i\theta}}{2} \tag{1.39}$$

$$\sin\theta = \frac{e^{i\theta} - e^{-i\theta}}{2i} \tag{1.40}$$

These relationships allow us to write a complex number in *complex exponential form* or more commonly *polar form*. This is given by

$$z = re^{i\theta} \tag{1.41}$$

The polar form can be very useful for calculation, since exponentials are so simple to work with. For example, the product of two complex numbers $z = re^{i\theta}$ and $w = \rho e^{i\phi}$ is given by

$$zw = (re^{i\theta})(\rho e^{i\phi}) = r\rho e^{i(\theta+\phi)} \tag{1.42}$$

Notice that *moduli multiply* and *arguments add.* Division is also very simple:

$$\frac{z}{w} = \frac{re^{i\theta}}{\rho e^{i\phi}} = \frac{r}{\rho} e^{i\theta} e^{i\phi} = \frac{r}{\rho} e^{i(\theta - \phi)} \tag{1.43}$$

The reciprocal of a complex number takes on the relatively simple form:

$$z = re^{i\theta} \Rightarrow z^{-1} = \frac{1}{r} e^{-i\theta} \tag{1.44}$$

Raising a complex number to a power is also easy:

$$z^n = (re^{i\theta})^n = r^n e^{in\theta} \tag{1.45}$$

The complex conjugate is just

$$\overline{z} = re^{-i\theta} \tag{1.46}$$

Euler's formula can be used to derive some interesting expressions. For example, we can easily derive one of the most mysterious equations in all of mathematics:

$$\begin{aligned} e^{i\pi} &= \cos\pi + i\sin\pi \\ \Rightarrow e^{i\pi} &+ 1 = 0 \end{aligned} \tag{1.47}$$

DE MOIVRE'S THEOREM

Let $z_1 = r_1(\cos\theta_1 + i\sin\theta_1)$ and $z_2 = r_2(\cos\theta_2 + i\sin\theta_2)$. Using trigonometric identities and some algebra we can show that

$$z_1 z_2 = r_1 r_2[\cos(\theta_1 + \theta_2) + i\sin(\theta_1 + \theta_2)] \tag{1.48}$$

$$z_1 / z_2 = \frac{r_1}{r_2}[\cos(\theta_1 - \theta_2) + i\sin(\theta_1 - \theta_2)] \tag{1.49}$$

$$z_1 z_2 \ldots z_n = r_1 r_2 \ldots r_n[\cos(\theta_1 + \theta_2 + \cdots + \theta_n) + i\sin(\theta_1 + \theta_2 + \cdots + \theta_n)] \tag{1.50}$$

De Moivre's formula follows:

$$z^n = [r(\cos\theta + i\sin\theta)]^n = r^n(\cos n\theta + i\sin n\theta) \tag{1.51}$$

The *n*th Roots of Unity

Consider the equation

$$z^n = 1$$

where n is a positive integer. This innocuous looking equation actually has a bit of hidden data in it, this comes from the fact that

$$(e^z)^n = e^z e^z \cdots e^z$$

The *nth roots of unity* are given by

$$\cos 2k\pi / n + i\sin 2k\pi / n = e^{2k\pi i/n} \quad k = 0,1,2,\ldots,n-1 \tag{1.52}$$

If $w = e^{2\pi i/n}$ then the n roots are $1, w, w^2, \ldots, w^{n-1}$.

EXAMPLE 1.7
Show that $\cos z = \cos x \cosh y - i\sin x \sinh y$.

SOLUTION
This can be done using Euler's formula:

$$\begin{aligned}\cos(x+iy) &= \frac{e^{i(x+iy)} + e^{-i(x+iy)}}{2} \\ &= \frac{e^{ix-y} + e^{-ix+y}}{2} \\ &= \frac{e^{ix-y} + e^{-ix+y} + e^{ix-y} + e^{-ix+y}}{4}\end{aligned}$$

Now we can add and subtract some desired terms:

$$\frac{e^{ix-y}+e^{-ix+y}+e^{ix-y}+e^{-ix+y}}{4}$$
$$=\frac{e^{ix+y}+e^{-ix+y}+e^{ix-y}+e^{-ix-y}-e^{ix+y}+e^{-ix+y}+e^{-ix+y}-e^{-ix-y}}{4}$$
$$=\frac{e^{ix+y}+e^{-ix+y}+e^{ix-y}+e^{-ix-y}}{4}-\frac{e^{ix+y}-e^{-ix+y}-e^{-ix+y}+e^{-ix-y}}{4}$$
$$=\left(\frac{e^{ix}+e^{-ix}}{2}\right)\left(\frac{e^{y}+e^{-y}}{2}\right)-\left(\frac{e^{ix}-e^{-ix}}{2}\right)\left(\frac{e^{y}-e^{-y}}{2}\right)$$
$$=\left(\frac{e^{ix}+e^{-ix}}{2}\right)\left(\frac{e^{y}+e^{-y}}{2}\right)-i\left(\frac{e^{ix}-e^{-ix}}{2i}\right)\left(\frac{e^{y}-e^{-y}}{2}\right)$$
$$=\cos x\cosh y-i\sin x\sinh y$$

EXAMPLE 1.8

Show that $\sin^{-1} z=-i\ln(iz\pm\sqrt{1-z^2})$.

SOLUTION

We start with the relation

$$\cos^2\theta+\sin^2\theta=1$$

This means that we can write

$$\cos\theta=\pm\sqrt{1-\sin^2\theta}$$

Now let $\theta=\sin^{-1} z$. Then we have

$$\cos(\sin^{-1} z)=\pm\sqrt{1-\sin^2(\sin^{-1} z)}=\pm\sqrt{1-z^2}$$

This is true because $\sin(\sin^{-1}(\phi))=\phi$. Now we turn to Euler's formula:

$$e^{i\theta}=\cos\theta+i\sin\theta$$

Again, setting $\theta = \sin^{-1} z$ we have

$$e^{i\sin^{-1} z} = \cos(\sin^{-1} z) + i\sin(\sin^{-1} z)$$
$$= \pm\sqrt{1-z^2} + iz$$

Taking the natural logarithm of both sides, we obtain the desired result:

$$i\sin^{-1} z = \ln(iz \pm \sqrt{1-z^2})$$
$$\Rightarrow \sin^{-1} z = -i\ln(iz \pm \sqrt{1-z^2})$$

EXAMPLE 1.9

Show that $e^{\ln z} = re^{i\theta}$.

SOLUTION

We use the fact that $\theta = \theta + 2n\pi$ for $n = 0, 1, 2, \ldots$ to get

$$e^{\ln z} = e^{\ln(re^{i\theta})}$$
$$= e^{\ln r + \ln(e^{i\theta})}$$
$$= e^{\ln r + i\theta}$$
$$= e^{\ln r + i(\theta + 2n\pi)}$$
$$= re^{i\theta}e^{i2n\pi}$$
$$= re^{i\theta}(\cos 2n\pi + i\sin 2n\pi) = re^{i\theta}$$

EXAMPLE 1.10

Find the fourth roots of 2.

SOLUTION

We find the nth roots of a number a by writing $r^n e^{in\theta} = a e^{i0}$ and equating moduli and arguments, and repeating the process by adding 2π. This may not be clear, but we'll show this with the current example. First we start out with

$$(re^{i\theta})^4 = 2e^{i0}$$
$$\Rightarrow r = 2^{1/4} \qquad \theta = 0$$

This is the first of four roots. The second root is

$$(re^{i\theta})^4 = r^4 e^{i4\theta} = 2e^{i2\pi}$$
$$\Rightarrow r = 2^{1/4} \qquad \theta = \frac{\pi}{2}$$

So the second root is $z = 2^{1/4} e^{i\pi/2} = 2^{1/4}[\cos(\pi/2) + i\sin(\pi/2)] = i2^{1/4}$. Next, we have

$$(re^{i\theta})^4 = r^4 e^{i4\theta} = 2e^{i4\pi}$$
$$\Rightarrow r = 2^{1/4} \qquad \theta = \pi$$

And the root is

$$z = 2^{1/4} e^{i\pi} = 2^{1/4}(\cos\pi + i\sin\pi) = -2^{1/4}$$

The fourth and final root is found using

$$(re^{i\theta})^4 = r^4 e^{i4\theta} = 2e^{i6\pi}$$
$$\Rightarrow r = 2^{1/4} \qquad \theta = \frac{3\pi}{2}$$

In Cartesian form, the root is

$$z = 2^{1/4} e^{i\pi} = 2^{1/4}\left(\cos\left(\frac{3\pi}{2}\right) + i\sin\left(\frac{3\pi}{2}\right)\right) = -i2^{1/4}$$

Summary

The imaginary unit $i = \sqrt{-1}$ can be used to solve equations like $x^2 + 1 = 0$. By denoting real and imaginary parts, we can construct complex numbers that we can add, subtract, multiply, and divide. Like the reals, the complex numbers form a field. These notions can be abstracted to complex variables, which can be written in Cartesian or polar form.

Quiz

1. What is the modulus of $z = \dfrac{1-i}{4}$?

2. Write $z = \dfrac{2-4i}{3+2i-i^5}$ in standard form $z = x + iy$.

3. Find the sum and product of $z = 2 + 3i, w = 3 - i$.
4. Write down the complex conjugates of $z = 2 + 3i, w = 3 - i$.
5. Find the principal argument of $\frac{i}{-2-2i}$.
6. Using De Moivre's formula, what is $\sin 3\theta$?
7. Following the procedure outlined in Example 1.7, find an expression for $\sin(x + iy)$.
8. Express $\cos^{-1} z$ in terms of the natural logarithm.
9. Find all of the cube roots of i.
10. If $z = 16e^{i\pi}$ and $w = 2e^{i\pi/2}$, what is $\frac{z}{w}$?

CHAPTER 2

Functions, Limits, and Continuity

In the last chapter, although we saw a couple of functions with complex argument z, we spent most of our time talking about complex numbers. Now we will introduce complex *functions* and begin to introduce concepts from the study of calculus like limits and continuity. Many important points in the first few chapters will be covered several times, so don't worry if you don't understand everything right away.

Complex Functions

When we write z, we are denoting a complex variable, which is a symbol that can take on any value of a complex number. This is the same concept you are used to from real variables where we use x or y to represent a variable. We define a *function* of a complex variable $w = f(z)$ as a rule that assigns to each $z \in \mathbb{C}$ a complex

number w. If the function is defined only over a restricted set S, then $w = f(z)$ assigns to each $z \in S$ the complex number w and we call S the *domain* of the function. The *value* of a function at $z = a$ is indicated by writing $f(a)$.

EXAMPLE 2.1

Consider the function $f(z) = z^3$ and consider its value at $z = i$, $z = 1 + i$.

SOLUTION

In the first case we set $z = i$ and so we have

$$f(i) = i^3 = i(i^2) = -i$$

Now we let $z = 1 + i$. Since

$$z^2 = (1+i)^2 = (1+i)(1+i) = 1 + 2i - 1 = 2i$$

The value of the function is

$$f(1+i) = (1+i)^3 = (1+i)^2(1+i) = 2i(1+i) = -2 + 2i$$

A complex function can be a function of the complex conjugate $\bar{z}$ as well, so we could write $f(z, \bar{z})$. That is, we are treating z and its conjugate $\bar{z}$ as independent variables the way we might for a function of x and y, $g(x, y)$. Just as we could compute partial derivatives $\partial g/\partial x$ and $\partial g/\partial y$ to determine how g depends on x and y, we can determine how a complex function depends on z and its conjugate $\bar{z}$ by computing partial derivatives of the function with respect to each of these variables. As we will see in the next chapter, functions that do not depend on $\bar{z}$ have important properties, and in fact the study of complex analysis is the study of functions for which $\partial f/\partial \bar{z} = 0$.

EXAMPLE 2.2

Suppose that $f(z) = z^2\bar{z}$. Find $f(1+i)$

SOLUTION

In Example 2.1, we saw that $z^2 = (1+i)^2 = 2i$. The complex conjugate of $z = 1 + i$ is given by

$$\bar{z} = 1 - i$$

So we have

$$f(1+i) = z^2\bar{z} = (1+i)^2(1-i) = 2i(1-i) = 2+2i$$

The domain of a function can be restricted to a region where the function is well behaved. For example, the function

$$f(z) = \frac{1}{z}$$

is not defined at the origin. Let S be a domain that includes a region where the function is defined. Suppose that D is a region where the function is *not* defined. We can indicate that we are excluding a certain set from the domain of the function using the notation

$$z \in S \setminus D \tag{2.1}$$

For example, letting $f(z) = 1/z$, we see that the function is defined throughout the complex plane *except* at the origin. We can indicate this by writing

$$z \in \mathbb{C} \setminus \{0\} \tag{2.2}$$

Simply put, the domain of a function is a region where the function does not blow up.

EXAMPLE 2.3

What is the domain of definition for $f(z) = 1/(1+z^2)$.

SOLUTION

We can factorize the denominator and write the function in the following way:

$$f(z) = \frac{1}{1+z^2} = \frac{1}{(i+z)(-i+z)}$$

We can see that the function goes to infinity if

$$z = \pm i$$

Therefore the function is defined throughout the complex plane except at the points $z = \pm i$.

Something we'll repeatedly emphasize in the next couple of chapters is that a function, just like a complex number, can be written in terms of real and imaginary parts. Recall that the complex variable z can be written as $z = x + iy$. We call x the real part of z and y the imaginary part of z, but both x and y are themselves real variables. This concept carries over to a complex function, which can be written in the form $f(z) = f(x+iy) = u(x,y) + iv(x,y)$. The real part of f is given by

$$\text{Re}(f) = u(x,y) \tag{2.3}$$

And the imaginary part of f is given by

$$\text{Im}(f) = v(x,y) \tag{2.4}$$

Notice that we can write down the complex conjugate of a function. With $f(z) = f(x+iy) = u(x,y) + iv(x,y)$ the complex conjugate is given by

$$\overline{f}(z) = \overline{f}(x+iy) = u(x,y) - iv(x,y) \tag{2.5}$$

The same rule applied to complex numbers and complex variables was used, namely, we let $i \to -i$ in order to obtain the complex conjugate. Note that $u(x,y)$ and $v(x,y)$ are unchanged by this operation because they are both *real functions* of the *real variables* x and y.

In chap. 1, we learned how to write the real and imaginary parts of z in terms of z, $\overline{z}$ using Eqs. (1.11) and (1.12). We can write down analogous formulas for the real and imaginary parts of a function. First let's consider the real part of a complex function. We can add the function to its complex conjugate

$$f + \overline{f} = u + iv + u - iv = 2u$$

Hence the real part of a complex function is given by

$$u(x,y) = \frac{f(z) + \overline{f(z)}}{2} \tag{2.6}$$

And we can take the difference between a function and its complex conjugate:

$$f - \overline{f} = u + iv - (u - iv) = 2iv$$

This gives us the imaginary part of a complex function:

$$v(x,y) = \frac{f(z) - \overline{f(z)}}{2i} \tag{2.7}$$

EXAMPLE 2.4

What is the complex conjugate of $f(z) = 1/z$.

SOLUTION

First of all we can write the function as

$$f(z) = f(x+iy) = \frac{1}{x+iy}$$

Therefore the complex conjugate is

$$\bar{f} = \overline{\left(\frac{1}{x+iy}\right)} = \frac{1}{x-iy} = \frac{1}{\bar{z}}$$

EXAMPLE 2.5

What are the real and imaginary parts of $f(z) = z + (1/z)$.

SOLUTION

We let $z = x + iy$. Then we have

$$f(x+iy) = x + iy + \frac{1}{x+iy}$$

We need to write the second term in standard Cartesian notation. This is done by multiplying and dividing by its complex conjugate:

$$\frac{1}{x+iy} = \frac{1}{x+iy}\left(\frac{x-iy}{x-iy}\right) = \frac{x-iy}{x^2+y^2}$$

So, we have

$$\begin{aligned} f(x+iy) &= x + iy + \frac{x-iy}{x^2+y^2} \\ &= x + \frac{x}{x^2+y^2} + iy - \frac{iy}{x^2+y^2} \\ &= \frac{x^3+xy^2+x}{x^2+y^2} + i\left(\frac{y^3+x^2y-y}{x^2+y^2}\right) \end{aligned}$$

So the real part of the function is

$$\mathrm{Re}(f) = \frac{x^3+xy^2+x}{x^2+y^2} = u(x,y)$$

The imaginary part of the function is

$$\mathrm{Im}(f) = \frac{y^3 + x^2 y - y}{x^2 + y^2} = v(x, y)$$

Note that we can write the real and imaginary parts in terms of $z, \overline{z}$ as follows. We have

$$\overline{f(z)} = \overline{z + \frac{1}{z}} = \overline{z} + \frac{1}{\overline{z}}$$

Now

$$\begin{aligned} f + \overline{f} &= z + \frac{1}{z} + \overline{z} + \frac{1}{\overline{z}} = z + \overline{z} + \frac{\overline{z}}{z\overline{z}} + \frac{z}{z\overline{z}} \\ &= \left(\frac{z + \overline{z}}{z\overline{z}}\right) z\overline{z} + \frac{z + \overline{z}}{z\overline{z}} \\ &= \frac{z^2\overline{z} + z\overline{z}^2 + z + \overline{z}}{z\overline{z}} \end{aligned}$$

So

$$\mathrm{Re}(f) = \frac{f + \overline{f}}{2} = \frac{z^2\overline{z} + z\overline{z}^2 + z + \overline{z}}{2z\overline{z}}$$

To get the imaginary part we calculate

$$\begin{aligned} f - \overline{f} &= z + \frac{1}{z} - \left(\overline{z} + \frac{1}{\overline{z}}\right) = z + \overline{z} - \frac{\overline{z}}{z\overline{z}} - \frac{z}{z\overline{z}} \\ &= \left(\frac{z + \overline{z}}{z\overline{z}}\right) z\overline{z} - \left(\frac{z + \overline{z}}{z\overline{z}}\right) \\ &= \frac{z^2\overline{z} + z\overline{z}^2 - z - \overline{z}}{z\overline{z}} \end{aligned}$$

Therefore

$$\mathrm{Im}(f) = \frac{f - \overline{f}}{2i} = \frac{z^2\overline{z} + z\overline{z}^2 - z - \overline{z}}{2iz\overline{z}}$$

In chap. 1 we also learned that a complex number $z = x + iy$ can be written in the polar representation $z = re^{i\theta}$. The same is true with complex functions. That is, we can write

$$f(z) = f(re^{i\theta}) \tag{2.8}$$

The function can also be written in terms of real and imaginary parts that are functions of the real variables r and θ. This is done as follows:

$$f(re^{i\theta}) = u(r,\theta) + iv(r,\theta) \tag{2.9}$$

EXAMPLE 2.6

Write the function $f(z) = z + (1/z)$ in the polar representation. What are the real and imaginary parts of the function?

SOLUTION

We write the function in the polar representation by letting $z = re^{i\theta}$. This gives

$$\begin{aligned} f(z) &= z + \frac{1}{z} \\ &= re^{i\theta} + \frac{1}{re^{i\theta}} \\ &= re^{i\theta} + \frac{1}{r}e^{-i\theta} \end{aligned}$$

Recalling Euler's formula, we can write $e^{\pm i\theta} = \cos\theta \pm i\sin\theta$. So the function becomes

$$\begin{aligned} f &= re^{i\theta} + \frac{1}{r}e^{-i\theta} \\ &= r(\cos\theta + i\sin\theta) + \frac{1}{r}(\cos\theta - i\sin\theta) \\ &= \cos\theta\left(r + \frac{1}{r}\right) + i\sin\theta\left(r - \frac{1}{r}\right) \\ &= \frac{\cos\theta}{r}(r^2 + 1) + i\frac{\sin\theta}{r}(r^2 - 1) \end{aligned}$$

This allows us to identify the real part of the function as

$$u(r,\theta) = \frac{\cos\theta}{r}(r^2 + 1)$$

The imaginary part of the function is given by

$$v(r,\theta) = \frac{\sin\theta}{r}(r^2 - 1)$$

Note that both the real and imaginary parts of the function are *real* functions of the *real* variables (r,θ).

Plotting Complex Functions

One of the most useful tools in the study of real functions is the ability to graph or plot them. This lets us get a feel for the functions behavior, for example we can see how it behaves as x gets large or look for points of discontinuity.

Unfortunately, in the case of a complex function we can't just plot the function the way we would a real function $f(x)$ of the real variable x. But, there are a few things we can look at. We can plot

- The real part of $f(z)$
- The imaginary part of $f(z)$
- The modulus or absolute value $|f(z)|$

In addition, if the function is written in polar representation, we can plot the argument of the function $\arg(f(z))$. We can also make a level set or contour plots of these items, or can plot them for a fixed point on the real or imaginary axis.

EXAMPLE 2.7

Plot the real and imaginary parts of $f(z) = z + (1/z)$.

SOLUTION

In Example 2.5 we found that

$$\mathrm{Re}(f) = \frac{x^3 + xy^2 + x}{x^2 + y^2} = u(x, y)$$

and

$$\mathrm{Im}(f) = \frac{y^3 + x^2y - y}{x^2 + y^2} = v(x, y)$$

Note that there is a singularity at the origin. A plot of the real part of the function is shown in Fig. 2.1. The imaginary part of the function has a similar form, as shown in Fig. 2.2. At the origin, the imaginary part of the function also blows up.

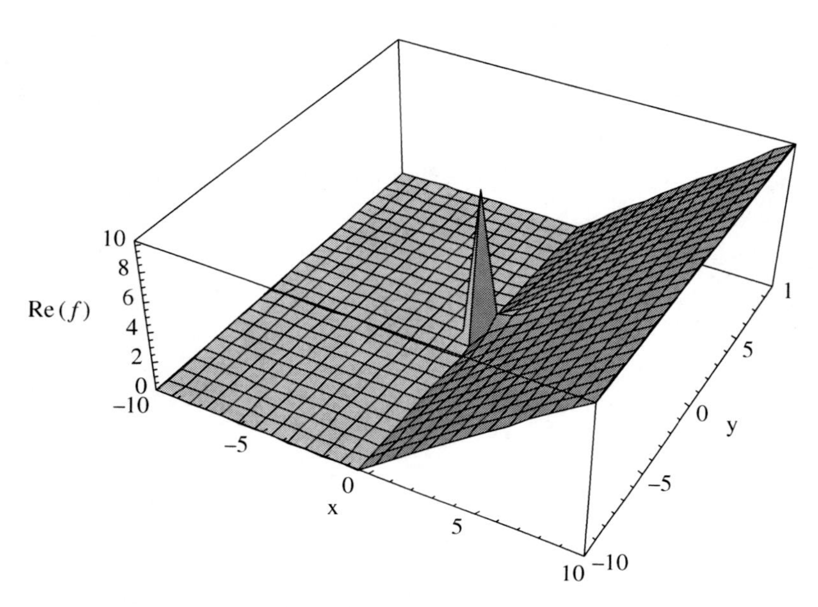

Figure 2.1 A plot of $\text{Re}(f) = (x^3 + xy^2 + x)/(x^2 + y^2) = u(x, y)$. The spike at the origin is a point where the function blows up.

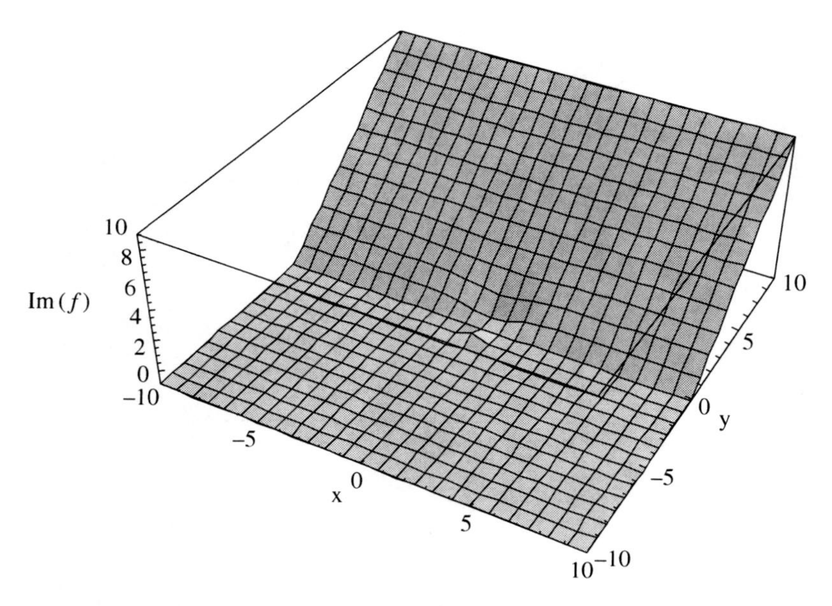

Figure 2.2 A plot of $\text{Im}(f) = (y^3 + x^2y - y)/(x^2 + y^2) = v(x, y)$.

EXAMPLE 2.8

Plot the absolute value of the function $f(z) = z + (1/z)$.

SOLUTION

Let's write down the absolute value of the function. It is given by

$$|f(z)| = f(z)\overline{f(z)}$$

So we have

$$\begin{aligned} f(z)\overline{f(z)} &= \left(z + \frac{1}{z}\right)\left(\overline{z} + \frac{1}{\overline{z}}\right) \\ &= z\overline{z} + \frac{z}{\overline{z}} + \frac{\overline{z}}{z} + \frac{1}{z\overline{z}} \\ &= \frac{z^2\overline{z}^2}{z\overline{z}} + \frac{z^2 + \overline{z}^2}{z\overline{z}} + \frac{1}{z\overline{z}} \\ &= \frac{z^2\overline{z}^2 + z^2 + \overline{z}^2 + 1}{z\overline{z}} \end{aligned}$$

Now we write this in terms of x and y:

$$\begin{aligned} |f(z)|^2 &= \frac{z^2\overline{z}^2 + z^2 + \overline{z}^2 + 1}{z\overline{z}} \\ &= \frac{(x+iy)^2(x-iy)^2 + (x+iy)^2 + (x-iy)^2 + 1}{x^2 + y^2} \\ &= \frac{(x^2 - y^2 + 2ixy)(x^2 - y^2 - 2ixy) + 2x^2 - 2y^2 + 1}{x^2 + y^2} \\ &= \frac{x^4 + 2x^2y^2 + y^4 + 2x^2 - 2y^2 + 1}{x^2 + y^2} \end{aligned}$$

Notice that this function will blow up at the origin, where $x = y = 0$. The absolute value is the square root:

$$|f(z)| = \sqrt{\frac{x^4 + 2x^2y^2 + y^4 + 2x^2 - 2y^2 + 1}{x^2 + y^2}}$$

A plot of this function is shown in Fig. 2.3.

EXAMPLE 2.9

Plot the real part of $f(z) = z + (1/z)$ along the line $x + i2$, for $-10 \le x \le 10$.

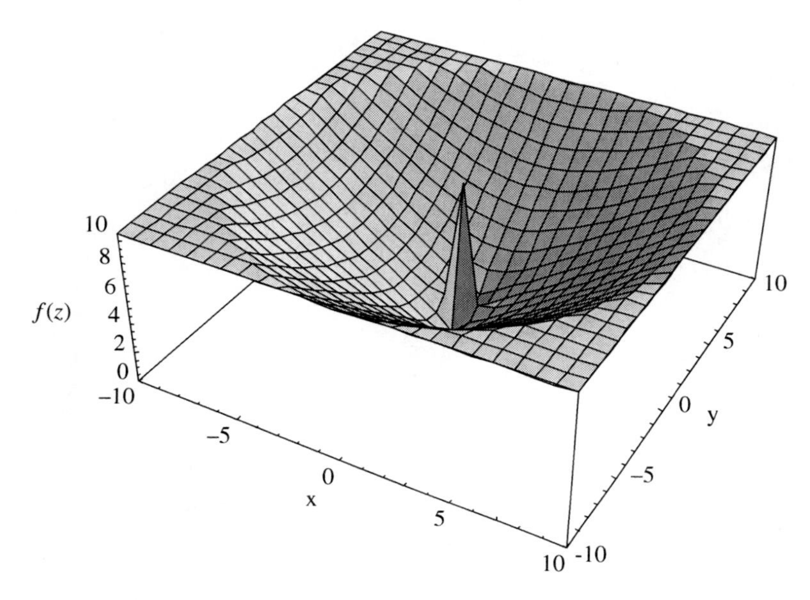

Figure 2.3 A plot of $|f(z)|$ for $f(z) = z + (1/z)$.

SOLUTION

Plotting with the real or imaginary part fixed like this is another way to study the behavior of the function. The real part is given by

$$\text{Re}(f) = \frac{x^3 + xy^2 + x}{x^2 + y^2} = u(x, y)$$

Setting $y = 2$ gives

$$u(x, 2) = \frac{x^3 + 5x}{x^2 + 4}$$

A plot of this function is shown in Fig. 2.4.

EXAMPLE 2.10

Generate a contour plot of $f(z) = 1/z$.

SOLUTION

A contour plot is a good way of visualizing where a function is increasing, decreasing, or blowing up. We show a contour plot of $|f(z)| = |1/z|$ in Fig. 2.5 generated with computer software. The plot shows increasing values in lighter

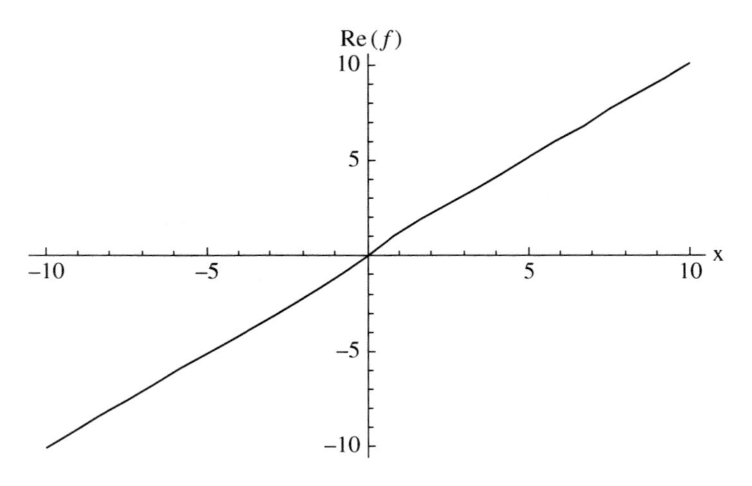

Figure 2.4 A plot of $\text{Re}(f) = (x^3 + xy^2 + x)/(x^2 + y^2) = u(x, y)$ with $y = 2$.

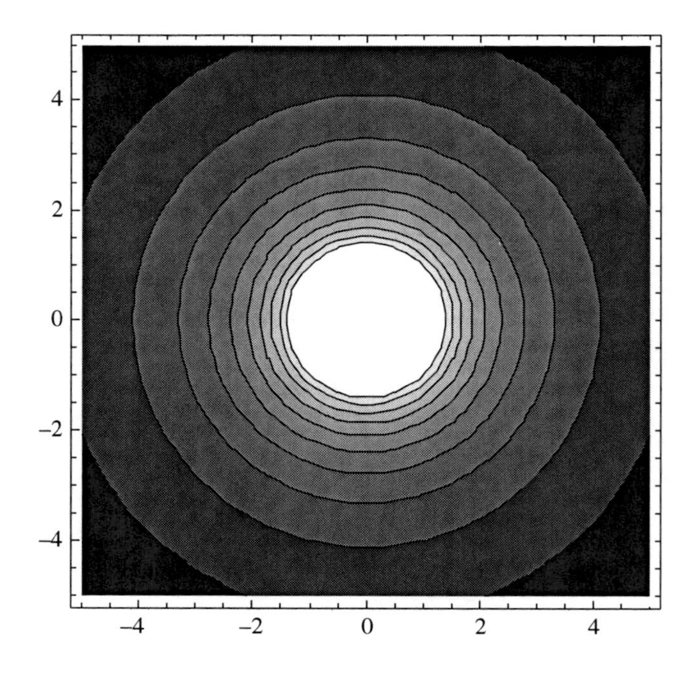

Figure 2.5 A contour plot of $|f(z)| = |1/z|$, showing zones where the function is increasing in magnitude and the point where it blows up at the origin.

colors—note the area about and including the origin is white indicating that the function blows up there.

Multivalued Functions

In many cases that we encounter in the theory of complex variables a function is *multivalued.* This is due to the periodic nature of the cosine and sin functions, Euler's formula $e^{i\theta} = \cos\theta + i\sin\theta$ and the fact that we can write $z = re^{i\theta}$ in the polar representation.

We say that a complex function $f(z)$ is *single valued* if only one value of w corresponds to each value of z where $w = f(z)$. If more than one value of w corresponds to each value of z, we say that the function is multivalued. A classic example of a multivalued function in complex variables is $\ln z$, which we discuss in chap. 4.

Limits of Complex Functions

Our first foray into the application of calculus to functions of a complex variable comes with the study of limits. Consider a point in the complex plane $z = a$ and let $f(z)$ be defined and *single valued* in some neighborhood about a. The neighborhood may include the point a, or we may omit a in which case we say that the function is defined and single valued in a *deleted neighborhood* of a. The limit ℓ of $f(z)$ as $z \to a$ is written as

$$\lim_{z \to a} f(z) = \ell \tag{2.10}$$

Formally, what this means is that for any number $\varepsilon > 0$ we can find a $\delta > 0$ such that $|f(z) - a| < \varepsilon$ whenever $0 < |z - a| < \delta$. For the limit to exist, it must be independent of the direction in which we approach $z = a$. Note that a limit only exists if the limit is independent of the way that we approach the point in question, a point which is illustrated in Example 2.14.

Limits in the theory of complex variables satisfy the same properties that limits do in the real case. Specifically, let us define

$$\lim_{z \to a} f(z) = A \qquad \lim_{z \to a} f(z) = B$$

Then the following hold

$$\lim_{z\to a}\{f(z)+g(z)\}=\lim_{z\to a}f(z)+\lim_{z\to a}g(z)=A+B \tag{2.11}$$

$$\lim_{z\to a}\{f(z)-g(z)\}=\lim_{z\to a}f(z)-\lim_{z\to a}g(z)=A+B \tag{2.12}$$

$$\lim_{z\to a}\{f(z)g(z)\}=\left\{\lim_{z\to a}f(z)\right\}\left\{\lim_{z\to a}g(z)\right\}=AB \tag{2.13}$$

$$\lim_{z\to a}\frac{f(z)}{g(z)}=\frac{\lim_{z\to a}f(z)}{\lim_{z\to a}g(z)}=\frac{A}{B} \tag{2.14}$$

Property in Eq. (2.14) holds as long as $B\neq 0$.

Limits can be calculated in terms of real and imaginary parts. Let $f=u+iv$, $z=x+iy$, and $z_0=x_0+iy_0, w=u_0+iv_0$. Then

$$\lim_{z\to z_0}f(z)=w_0$$

If and only if

$$\lim_{\substack{x\to x_0\\ y\to 0}}u(x,y)=u_0 \qquad \lim_{\substack{x\to x_0\\ y\to 0}}v(x,y)=v_0 \tag{2.15}$$

OPEN DISKS

Frequently, in complex analysis we wish to consider a circular region in the complex plane. We call such a region a *disk*. Suppose that the radius of the disk is a. If the points on the edge of the disk, that is, the points lying on the circular curve defining the border of the disk are not included in the region of consideration, we say that the disk is *open.*

Consider a disk of radius one centered at the origin. We indicate this by writing

$$|z|<1$$

This is illustrated in Fig. 2.6.

If the disk of radius r is instead centered at a point a, then we would write

$$|z-a|<r$$

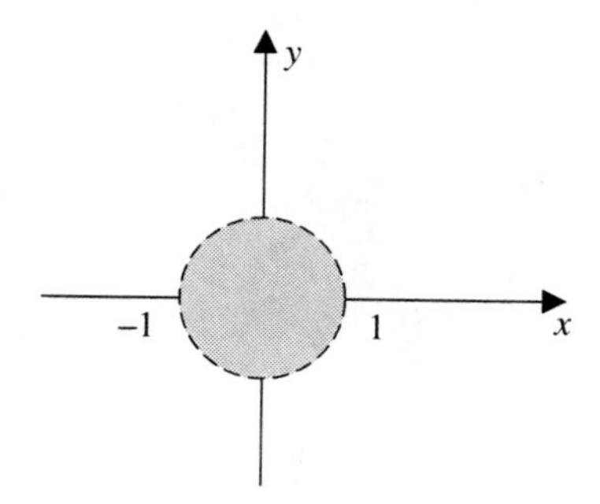

Figure 2.6 The disk $|z| < 1$ is centered at the origin. The boundary is indicated with a dashed line, which is sometimes done to indicate it is not included in the region of definition.

For example:

$$|z-3| < 5$$

describes a disk of radius five centered at the point $z = 3$. This is shown in Fig. 2.7.

EXAMPLE 2.11

Compute $\lim_{z\to 3}(iz-1)/2$ in the open disk $|z| < 3$.

SOLUTION

Notice that the point $z = 3$ is on the boundary of the domain of the function. Just plugging in we find

$$\lim_{z\to 3}\frac{iz-1}{2} = -\frac{1}{2} + i\frac{3}{2}$$

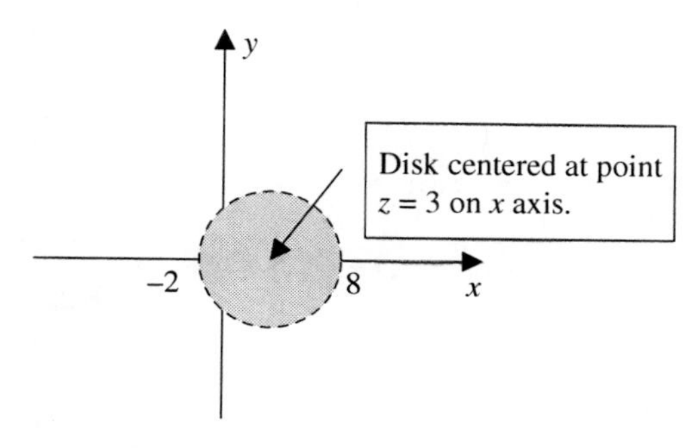

Figure 2.7 The disk described by $|z-3| < 5$.

Let's confirm this by applying the formal definition of the limit. Notice that

$$\begin{aligned}|f(z)-\ell| &= \left|\frac{iz-1}{2}\right| = \left|\frac{iz-1}{2}+\frac{1}{2}-i\frac{3}{2}\right| \\ &= \left|\frac{i}{2}(z-3)\right| \\ &= \frac{|z-3|}{2}\end{aligned}$$

So we've found that given any $\varepsilon > 0$

$$\left|f(z)-\left(\frac{1}{2}+i\frac{3}{2}\right)\right| < \varepsilon$$

whenever

$$0 < |z-3| < 2\varepsilon$$

EXAMPLE 2.12
Compute $\lim_{z\to i}(z^2)(z+i)$.

SOLUTION
For illustration purposes, we compute the limits of z^2 and $z+i$ independently and then apply Eq. (2.13). First we have

$$\lim_{z\to i} z^2 = i^2 = -1$$

Secondly

$$\lim_{z\to i} z+i = i+i = 2i$$

Hence

$$\lim_{z\to i}(z^2)(z+i) = (-1)(2i) = -2i$$

EXAMPLE 2.13
Using the theorems on limits from Eqs. (2.11)–(2.14) evaluate $\lim_{z\to 2i} f(z)$ when $f(z) = z^2 + 2z + 5$.

SOLUTION

We have

$$\begin{aligned}\lim_{z\to 2i} z^2 + 2z + 5 &= \lim_{z\to 2i} z^2 + \lim_{z\to 2i} 2z + \lim_{z\to 2i} 5 \\ &= \left(\lim_{z\to 2i} z\right)\left(\lim_{z\to 2i} z\right) + 2\lim_{z\to 2i} z + \lim_{z\to 2i} 5 \\ &= (2i)(2i) + 2(2i) + 5 \\ &= -4 + 4i + 5 \\ &= 1 + 4i\end{aligned}$$

EXAMPLE 2.14

Show that the limit $\lim_{z\to 0} \bar{z}/z$ does not exist.

SOLUTION

A limit only exists if the limit is independent of the way that we approach the point in question. For this limit, let's calculate it in two different ways. The first way we'll calculate it is by approaching the origin *along the x axis*. This means that we set $y = 0$, so

$$\frac{\bar{z}}{z} = \frac{x - iy}{x + iy} \underset{y=0}{\rightarrow} \frac{x}{x} = 1$$

Hence

$$\lim_{z\to 0} \frac{\bar{z}}{z} = +1$$

Now, instead we choose to approach the origin along the y axis. This means that we will have to set $x = 0$. So we obtain

$$\frac{\bar{z}}{z} = \frac{x - iy}{x + iy} \underset{x=0}{\rightarrow} \frac{-iy}{iy} = -1$$

That is

$$\lim_{z\to 0} \frac{\bar{z}}{z} = -1$$

Since we obtained two different values for the limit by approaching the origin in two different directions, the limit cannot exist.

Limits Involving Infinity

A limit $\lim_{z \to a} f(z)$ blows up or goes to infinity $\lim_{z \to a} f(z) = \infty$ if and only if

$$\lim_{z \to a} \frac{1}{f(z)} = 0 \tag{2.16}$$

The limit as z goes to infinity is equal to ℓ if and only if

$$\lim_{z \to 0} f\left(\frac{1}{z}\right) = \ell \tag{2.17}$$

If Eq. (2.17) holds, then we can write $\lim_{z \to \infty} f(z) = \ell$. Finally, $\lim_{z \to \infty} f(z) = \infty$ if and only if

$$\lim_{z \to 0} \frac{1}{f(1/z)} = 0 \tag{2.18}$$

EXAMPLE 2.15

Show that

$$\lim_{z \to -2} \frac{z+5}{z+2} = \infty$$

SOLUTION

We do this using Eq. (2.16). We have

$$\lim_{z \to -2} \frac{z+5}{z+2} \to \lim_{z \to -2} \frac{z+2}{z+5} = \frac{-2+2}{-2+5} = \frac{0}{3} = 0$$

Hence, $\lim_{z \to -2} (z+5)/(z+2) = \infty$.

Continuity

A function $f(z)$ is *continuous* at a point $z = a$ if the following three conditions are satisfied:

- $\lim_{z \to a} f(z)$ exists
- $f(a)$ exists
- $\lim_{z \to a} f(z) = f(a)$

EXAMPLE 2.16

Suppose that

$$f(z)=\begin{cases} z^2 & \text{for } z \neq i \\ 0 & \text{for } z = i \end{cases}$$

Show that the function is not continuous.

SOLUTION

We know intuitively that the function is not continuous since the value of the function changes suddenly at the point $z=i$. It is not too much work to show that the limit exists. We have

$$\left|z^2-a^2\right|=|z-i||z+i|<\delta|z-i+2i|<\delta\{|z-i|+|2i|\}<\delta\{1+2|i|\}=3\delta$$

Take δ equal to the minimum of $1, \frac{\varepsilon}{3}$ and then $\left|z^2-i^2\right|<\varepsilon$ whenever $|z-i|<\delta$. So the limit exists. In particular

$$\lim_{z\to i} z^2 = i^2 = -1$$

The function also meets the second condition, namely that it is defined at the point $z=i$:

$$f(i)=0$$

where the analysis fails in comparing the limit of the function as it approaches the point to its value at the point. In this case:

$$\lim_{z\to i} z^2 = i^2 = -1 \text{ and } f(i)=0$$
$$\Rightarrow \lim_{z\to a} f(z) \neq f(a)$$

This establishes in a formal sense what we already knew intuitively, that the function is not continuous at $z=i$.

EXAMPLE 2.17

Show that $f(z)=z^3$ is continuous at $z=i$.

SOLUTION

The function is continuous at $z=i$. We have

$$\lim_{z\to i} z^3 = i^3 = -i$$

Since

$$f(i) = i^3 = -i, \lim_{z \to i} z^3 = f(i)$$

and we conclude that the function is continuous at $z = i$. The limit can be evaluated using the ε, δ approach achieving the same results.

Summary

In this chapter, we introduced some notions that will allow us to develop a calculus of complex variables. Namely, we introduced the concept of a function. We indicated that a function of a complex variable can be single or multiple valued. In a region where a function is single valued, we demonstrated how to compute basic limits and explored an elementary notion of continuity. We also illustrated how plots of complex functions can be generated.

Quiz

1. Evaluate $f(1+i)$ when $f(z) = z^2 + i$.
2. Evaluate $f(1+i)$ when $f(z) = |z|^2 + i$.
3. Find $\overline{f(z)}$ when $f(z) = z^2 + 2z\overline{z} + i$.
4. What is the real part of $f = z + \overline{z}^2 + 3$?
5. Write $f = u(x,y) + iv(x,y)$ if $f = 1 - z + |z|^2$.
6. Find $\lim_{z \to 2+i} (z^2 + 1)$ for the real part of $z^2 + 1$.
7. Compute $\lim_{z \to \infty} \dfrac{z^2}{(z-1)^2}$.
8. Compute $\lim_{z \to i} \dfrac{iz^3 + 1}{z + i}$.
9. Find $\lim_{z \to 0} \dfrac{\overline{z}^2}{z}$.
10. Is $f(z) = \dfrac{3z^2 - 3}{z - 1}$ continuous?

CHAPTER 3

The Derivative and Analytic Functions

The next step in extending the calculus of real variables to include complex variables is to define the notion of a derivative. You won't be surprised to find out that computing derivatives or shall we say determining when a function is differentiable is a little more involved when considering functions of a complex variable. While we will see that many of the basic theorems about derivatives carry over from real to complex variables, there are some differences. In particular, we're going to have to pay attention to how we approach the origin when computing the limits used to define the derivative, and we'll find the unusual result that some functions of a complex variable are continuous but not differentiable. After learning this we'll see that functions of a complex variable that are differentiable satisfy a nice set of equations known as the *Cauchy-Riemann equations,* one of the most elegant results in pure and applied mathematics. Loosely speaking, a function which satisfies the Cauchy-Riemann equations is called *analytic*. Finally, we close the chapter with a look at a special type of function that satisfies *Laplace's equation* which we call a *harmonic function*. Harmonic functions are functions of the real variables x and y but they can be used to construct a complex function, which is analytic.

The Derivative Defined

Consider some point z_0 in the complex plane and let $f(z)$ be some function such that its domain contains a neighborhood of z_0. The derivative of $f(z)$ at the point z_0 is defined by the limit

$$f'(z_0) = \lim_{z \to z_0} \frac{f(z) - f(z_0)}{z - z_0} \tag{3.1}$$

If this limit exists for all points in a domain D, we say that $f(z)$ is differentiable in D. At the given point, if the limit exists we say that $f(z)$ is *complex differentiable* at the point z_0.

We can write this limit in a form that may be more familiar to you, considering what you learned in elementary calculus. First lets define $\Delta z = z - z_0$. Then the derivative of $f(z)$ at the point z_0 can be written as

$$f'(z_0) = \lim_{\Delta z \to 0} \frac{f(z_0 + \Delta z) - f(z_0)}{\Delta z} \tag{3.2}$$

In a moment, we'll be able to use this limit to write down the derivative in terms of Leibniz notation. Right now, we stop to make some important definitions.

Definition: Analytic and Entire Functions

Suppose that $f(z)$ is differentiable in an ε-neighborhood of the point z_0. That is, we define the domain D such that

$$|z - z_0| < \varepsilon$$

for some $\varepsilon > 0$. If $f'(z)$ exists for all $z \in D$ then we say that $f(z)$ is analytic at the point z_0.

As you might guess, many functions are differentiable everywhere—that is, throughout the entire complex plane. This turns out to be true for many but not all functions of a complex variable. However, if $f(z)$ is analytic on the whole complex plane then we say that the function $f(z)$ is *entire*.

Leibniz Notation

Now let's return to the definition of the derivative and consider Leibniz notation, which as you know from the calculus of real variables makes life a whole lot easier. To do this we make the notational definition $w = f(z)$. Then

$$\Delta w = f(z) - f(z_0)$$

With this notation together with the definition $\Delta z = z - z_0$ and, we have

$$f'(z_0) = \lim_{\Delta z \to 0} \frac{\Delta w}{\Delta z} = \frac{dw}{dz} \tag{3.3}$$

Let's explore the computation of derivatives with some examples.

EXAMPLE 3.1

Let $f(z) = z^2$ and find its derivative at any point z.

SOLUTION

Letting $w = z^2$ and using the definition given in we have

$$\lim_{\Delta z \to 0} \frac{(z + \Delta z)^2 - z^2}{\Delta z}$$

While it's really elementary in this specific case, we can expand the term $(z + \Delta z)^2$ using the *binomial theorem*. You should familiarize yourself with this technique so that you can handle more complicated cases. The binomial theorem tells us that

$$(x + y)^2 = x^2 + 2xy + y^2$$

Hence

$$(z + \Delta z)^2 = z^2 + 2z(\Delta z) + (\Delta z)^2$$

and so

$$\begin{aligned}
\lim_{\Delta z \to 0} \frac{(z + \Delta z)^2 - z^2}{\Delta z} &= \lim_{\Delta z \to 0} \frac{z^2 + 2z(\Delta z) + (\Delta z)^2 - z^2}{\Delta z} \\
&= \lim_{\Delta z \to 0} \frac{2z(\Delta z) + (\Delta z)^2}{\Delta z} \\
&= \lim_{\Delta z \to 0} 2z + \lim_{\Delta z \to 0} \Delta z \\
&= 2z
\end{aligned}$$

So we see that in the case of complex variables as in the calculus of a real variable

$$\frac{d}{dz}(z^2) = 2z$$

as expected. You might find this obvious, but it's comforting to know that what we learned in elementary calculus of real variables carries over to the complex case.

Unfortunately, everything that we learned in elementary calculus does not carry over. For example, consider a function that is continuous but *not* differentiable. Take $f(z) = \overline{z}$. Using the definition of the complex conjugate described in Chap. 1, we know that

$$f(z) = \overline{z} = \overline{x + iy} = x - iy$$

To see why this function is not differentiable, we consider approaching a point $z_0 = x_0 + iy_0$ in two different ways. If a function is differentiable, it will not matter how we approach the point. We should be able to approach $z_0 = x_0 + iy_0$ in two different ways and get the same value for the limit, which defines the derivative. In the case of $f(z) = \overline{z}$, things don't work out that way.

What we'll do in this case is approach the origin in two different ways. First we'll try it along the x axis so that we set $\Delta z = (\Delta x, 0)$. Then we will try it along the y axis, and in that case we'll set $\Delta z = (0, \Delta y)$. Then we will compare the results.

Now, notice that for $f(z) = \overline{z}$

$$\frac{\Delta w}{\Delta z} = \frac{\overline{z + \Delta z} - \overline{z}}{\Delta z} = \frac{\overline{z} + \overline{\Delta z} - \overline{z}}{\Delta z} = \frac{\overline{\Delta z}}{\Delta z}$$

If we set $\Delta z = (\Delta x, 0)$ then we have

$$\frac{\overline{z + \Delta z} - \overline{z}}{\Delta z} = \frac{\overline{x + iy + \Delta x} - (\overline{x + iy})}{\Delta x} = \frac{x - iy + \Delta x - x + iy}{\Delta x} = \frac{\Delta x}{\Delta x} = 1$$

That is, taking the limit along the x axis gives

$$\frac{dw}{dz} = 1$$

Now, let's instead consider approaching the origin along the y axis. Recall that this means we'll set $\Delta z = (0, \Delta y)$. Therefore

$$\frac{\overline{z + \Delta z} - \overline{z}}{\Delta z} = \frac{\overline{x + iy + i\Delta y} - (\overline{x + iy})}{i\Delta y} = \frac{x - iy - i\Delta y - x + iy}{i\Delta y} = -\frac{\Delta y}{\Delta y} = -1$$

This means that taking the limit along the y axis gives us

$$\frac{dw}{dz} = -1$$

We conclude that the derivative of $f(z) = \bar{z}$ does not exist, even though the function is continuous everywhere.

Rules for Differentiation

So far we've seen that the basic definition of the derivative used in calculus works with complex variables, but that not all functions of a complex variable are differentiable even if they are continuous. Remember, a function that is differentiable in some region D of the complex plane is called analytic. We'll see in the later section that using an elegant formulation called the Cauchy-Riemann equations makes it a simple matter to show whether or not a given function is analytic. What does this mean for us? We can dispense with having to examine pesky limits. Matters will simplify and we can just calculate derivatives like we would in elementary calculus.

When computing derivatives of a function of a complex variable, several key results carry over from the calculus of real variables. These include

- Rules for computing the derivative of a constant
- Rules for computing the derivative of a polynomial
- The product rule
- The quotient rule

Let's start by considering the derivative of a constant. Let α be a constant which is a complex number. Then

$$\frac{d}{dz}\alpha = 0 \tag{3.4}$$

It follows that if a constant multiplies some function $f(z)$, we can pass it right through a derivative operator

$$\frac{d}{dz}(\alpha f) = \alpha \frac{df}{dz} \tag{3.5}$$

The next we consider is the derivative of a polynomial. The rule used to compute the derivative of a polynomial in complex variables turns out to be the same as we use with real variables. We have already seen that $f'(z) = 2z$ when $f(z) = z^2$. Generally:

$$\frac{dz^n}{dz} = nz^{n-1} \tag{3.6}$$

Now let $f(z)$ and $g(z)$ be two complex functions. It is not hard to show that

$$\frac{d}{dz}(f \pm g) = \frac{df}{dz} \pm \frac{dg}{dz} \tag{3.7}$$

Combining this with Eqs. (3.5) and (3.6) we are able to write down the derivative of any polynomial. Later, we'll see an important by-product of this result. If we expand some complex function in a series:

$$f(z) = \sum_{n=0}^{\infty} a_n z^n$$

Then we can differentiate the function by differentiating the series term by term using what we already know:

$$\frac{df}{dz} = \frac{d}{dz}\left(\sum_{n=0}^{\infty} a_n z^n\right) = \sum_{n=1}^{\infty} n a_n z^{n-1}$$

When we study series in detail in a later chapter, note that we will need to consider the radius of convergence of the series.

EXAMPLE 3.2

Find the derivative of $f(z) = 5z^2 + 3z - 2$.

SOLUTION

This is an elementary problem we can solve by applying the rules for derivatives stated so far. We have

$$\begin{aligned}\frac{df}{dz} &= \frac{d}{dz}(5z^2 + 3z - 2) \\ &= \frac{d}{dz}(5z^2) + \frac{d}{dz}(3z) - \frac{d}{dz}(2) \\ &= 10z + 3\end{aligned}$$

Derivatives of Some Elementary Functions

The derivatives of many common functions like the exponential and trig functions follow from ordinary calculus as well. One way this can be understood is by noting that these functions can be expanded in a series and differentiated term by term. We can also momentarily return to the use of limits and compute the derivatives that way, obtaining many familiar results.

EXAMPLE 3.3

Find $f'(z)$ when $f(z) = e^z$.

SOLUTION

Using the definition of the derivative given in Eq. (3.2) we have

$$\begin{aligned}\frac{d}{dz}e^z &= \lim_{\Delta z \to 0} \frac{e^{z+\Delta z} - e^z}{\Delta z} \\ &= \lim_{\Delta z \to 0} \frac{e^z e^{\Delta z} - e^z}{\Delta z} \\ &= e^z \lim_{\Delta z \to 0} \frac{e^{\Delta z} - 1}{\Delta z}\end{aligned}$$

To proceed, we write down the real and imaginary parts explicitly. Recall Euler's formula $e^{i\theta} = \cos\theta + i\sin\theta$. This allows us to write $e^z = e^{x+iy} = e^x e^{iy} = e^x(\cos y + i\sin y)$. So, the limit can be written as

$$\begin{aligned}e^z \lim_{\Delta z \to 0} \frac{e^{\Delta z} - 1}{\Delta z} &= e^z \lim_{\Delta z \to 0} \frac{e^{\Delta x + i\Delta y} - 1}{\Delta x + i\Delta y} \\ &= e^z \lim_{\substack{\Delta x \to 0 \\ \Delta y \to 0}} \frac{e^{\Delta x}(\cos\Delta y + i\sin\Delta y) - 1}{\Delta x + i\Delta y} \\ &= e^z \lim_{\substack{\Delta x \to 0 \\ \Delta y \to 0}} \frac{e^{\Delta x}(\cos\Delta y - 1) + ie^{\Delta x}\sin\Delta y}{\Delta x + i\Delta y}\end{aligned}$$

Now, as $\Delta x \to 0$, $e^{\Delta x} \to 1$, and as $\Delta y \to 0$, $\cos\Delta y \to 1$. Hence the real part of the numerator goes as

$$e^{\Delta x}(\cos\Delta y - 1) \to 1(1-1) = 0$$

So let's concentrate on the imaginary part and set $\Delta x \to 0$. We will expand the sin function in Taylor, giving us

$$\lim_{\Delta y \to 0} \frac{\sin \Delta y}{\Delta y} = \lim_{\Delta y \to 0} \frac{\Delta y - (\Delta y)^3/3! + (\Delta y)^5/5! - \cdots}{\Delta y} = \lim_{\Delta y \to 0} \left(1 - \frac{(\Delta y)^2}{3!} + \frac{(\Delta y)^4}{5!} - \cdots\right) = 1$$

This means

$$\lim_{\Delta z \to 0} \frac{e^{\Delta z} - 1}{\Delta z} = 1$$

Therefore it must be the case that

$$\frac{d}{dz} e^z = \lim_{\Delta z \to 0} \frac{e^{z+\Delta z} - e^z}{\Delta z} = e^z$$

Other derivatives of elementary functions also correspond to the results you're familiar with from the calculus of real variables:

$$\begin{aligned} \frac{d}{dz}\sin z &= \cos z & \frac{d}{dz}\cos z &= -\sin z \\ \frac{d}{dz}\sinh z &= \cosh z & \frac{d}{dz}\cosh z &= \sinh z \end{aligned} \tag{3.8}$$

The Product and Quotient Rules

The product and quotient rules also carry over to the case of complex variables. We have

$$\frac{d}{dz}(fg) = \frac{df}{dz} g + \frac{dg}{dz} f \tag{3.9}$$

Provided that $g(z) \neq 0$:

$$\frac{d}{dz}\left(\frac{f}{g}\right) = \frac{f'g - g'f}{g^2} \tag{3.10}$$

Finally, we note the chain rule for composite functions. If $F(z) = g[f(z)]$ then

$$F'(z) = g'[f(z)]f'(z) \tag{3.11}$$

EXAMPLE 3.4

Find the derivatives of

$$F_1(z) = \frac{z+1}{2z+1} \qquad F_2(z) = (1-2z^2)^3$$

SOLUTION

In the first case, we use the quotient rule making the following identifications

$$f(z) = z+1 \qquad \Rightarrow f'(z) = 1$$
$$g(z) = 2z+1 \qquad \Rightarrow g'(z) = 2$$

Hence

$$\frac{f'g - g'f}{g^2} = \frac{(1)(2z+1)-(2)(z+1)}{(2z+1)^2} = -\frac{1}{(2z+1)^2}$$

Note that this result is valid provided that $2z+1 \neq 0$ or $z \neq -1/2$, otherwise the derivative would blow up.

Considering the second function, we can use the rule for the derivative of a composite function with

$$f(z) = 1-2z^2 \qquad \Rightarrow f'(z) = -4z$$
$$g(z) = f(z)^3 \qquad \Rightarrow g'(z) = 3f^2$$

And so:

$$F_2' = -12z(1-2z^2)^2$$

Before proceeding to the Cauchy-Riemann equations, we note two important theorems.

THEOREM 3.1

If $f(z)$ is differentiable at a point z_0, then it is also continuous at z_0.

PROOF

Writing out the definition of the derivative in terms of the limit, we have

$$f'(z_0) = \lim_{z\to z_0} \frac{f(z)-f(z_0)}{z-z_0}$$

Now, notice that

$$\lim_{z \to z_0} f(z) - f(z_0) = \lim_{z \to z_0} \frac{f(z) - f(z_0)}{z - z_0}(z - z_0)$$
$$= f'(z_0) \lim_{z \to z_0}(z - z_0) = 0$$

This means that

$$\lim_{z \to z_0} f(z) = f(z_0)$$

Hence it follows that if $f'(z)$ exists at z_0, $f(z)$ is continuous there.

THEOREM 3.2: L'HOPITAL'S RULE
Let $f(z)$ and $g(z)$ be two functions that are analytic at a point z_0. Then provided that $g'(z_0) \neq 0$, if $f(z_0) = g(z_0) = 0$ then

$$\lim_{z \to z_0} \frac{f(z)}{g(z)} = \lim_{z \to z_0} \frac{f'(z)}{g'(z)} \tag{3.12}$$

EXAMPLE 3.5
Find the following limit:

$$\lim_{z \to i} \frac{z - i}{z^2 - z + 1 + i}$$

SOLUTION
We see that $f(i) = i - i = 0$ and $g(i) = i^2 - i + 1 + i = -1 - i + 1 + i = 0$, so we apply the rule. Computing the derivatives we get

$$f'(z) = \frac{d}{dz}(z - i) = 1$$
$$g'(z) = \frac{d}{dz}(z^2 - z + 1 + i) = 2z - 1$$

Therefore

$$\lim_{z \to i} \frac{z - i}{z^2 - z + 1 + i} = \lim_{z \to i} \frac{1}{2z - 1} = \frac{1}{2i - 1}$$

The Cauchy-Riemann Equations

Now let's work our way up to one of the most important and elegant results in all of mathematics, the Cauchy-Riemann equations. These equations, which were independently discovered by the mathematicians Augustin Louis Cauchy (1789–1857) and George Friedrich Bernhard Riemann (1826–1866) (Riemann derived them in his doctoral dissertation) allow us to quickly determine whether or not a function is analytic. To start, we write a function $f(z)$ of a complex variable in terms of real and imaginary parts:

$$f(z) = u(x, y) + iv(x, y) \tag{3.13}$$

The real and imaginary parts are themselves functions, but they are real functions of the real variables x and y.

What we'll be after in determining whether or not a function is analytic is to find out how that function depends on z and $\overline{z}$. First we make a definition.

Definition: Continuously Differentiable Function

Consider an open region D in the complex plane and a function $f : D \to \mathbb{C}$. If this function is continuous and if the partial derivatives $\partial f / \partial x$ and $\partial f / \partial y$ exist and are continuous, we say that f is continuously differentiable in D. If f is k times continuously differentiable where $k = 0, 1, 2, \ldots$ (that is, k derivatives of f exist and are continuous) we say that f is C^k. If f is C^0, this is a continuous function which is not differentiable.

Now, how do we determine if a function is analytic? If f is a continuously differentiable function on some region D then it is analytic if it has *no dependence* on $\overline{z}$. That function is analytic in a domain D provided that

$$\frac{df}{d\overline{z}} = 0 \tag{3.14}$$

This condition must hold for all points in D. Note that a function which is analytic is also called *holomorphic*. Later we will see that we can form a local power series expansion of a holomorphic or analytic function.

Now let's go back to writing a function of a complex variable in terms of real and imaginary parts as in Eq. (3.13). We want to think about how to compute the derivative d/dz in terms of derivatives with respect to the real variables x and y. Let's go back to basics. Remember that $z = x + iy$. This tells us that

$$\frac{\partial z}{\partial x} = 1 \quad \text{and} \quad \frac{\partial z}{\partial y} = i \tag{3.15}$$

Since $\bar{z} = x - iy$, it is the case that

$$\frac{\partial \bar{z}}{\partial x} = 1 \qquad \text{and} \qquad \frac{\partial \bar{z}}{\partial y} = -i \tag{3.16}$$

These formulas can be inverted. Recalling from Chap. 1 that $x = (z + \bar{z})/2$ and $y = (z - \bar{z})/2i$ we find that

$$\frac{\partial x}{\partial z} = \frac{1}{2} = \frac{\partial x}{\partial \bar{z}} \tag{3.17}$$

and

$$\frac{\partial y}{\partial z} = -\frac{i}{2} \qquad \frac{\partial y}{\partial \bar{z}} = +\frac{i}{2} \tag{3.18}$$

(remember that $1/i = -i$.) Using these results we can write the derivatives $\partial/\partial z$ and $\partial/\partial \bar{z}$ in terms of the derivatives $\partial/\partial x$ and $\partial/\partial y$. In the first case we have

$$\frac{\partial}{\partial z} = \frac{\partial x}{\partial z}\frac{\partial}{\partial x} + \frac{\partial y}{\partial z}\frac{\partial}{\partial y} = \frac{1}{2}\frac{\partial}{\partial x} - \frac{i}{2}\frac{\partial}{\partial y} \tag{3.19}$$

Similarly

$$\frac{\partial}{\partial \bar{z}} = \frac{\partial x}{\partial \bar{z}}\frac{\partial}{\partial x} + \frac{\partial y}{\partial \bar{z}}\frac{\partial}{\partial y} = \frac{1}{2}\frac{\partial}{\partial x} + \frac{i}{2}\frac{\partial}{\partial y} \tag{3.20}$$

Now we can use these results to write the derivatives df/dz and $df/d\bar{z}$ in terms of derivatives with respect to the real variables x and y:

$$\begin{aligned} \frac{\partial f}{\partial z} &= \frac{1}{2}\left(\frac{\partial}{\partial x} - i\frac{\partial}{\partial y}\right) f = \frac{1}{2}\left(\frac{\partial}{\partial x} - i\frac{\partial}{\partial y}\right)(u + iv) \\ &= \frac{1}{2}\left(\frac{\partial u}{\partial x} + \frac{\partial v}{\partial y}\right) + \frac{i}{2}\left(\frac{\partial v}{\partial x} - \frac{\partial u}{\partial y}\right) \end{aligned} \tag{3.21}$$

$$\begin{aligned} \frac{\partial f}{\partial \bar{z}} &= \frac{1}{2}\left(\frac{\partial}{\partial x} + i\frac{\partial}{\partial y}\right) f = \frac{1}{2}\left(\frac{\partial}{\partial x} + i\frac{\partial}{\partial y}\right)(u + iv) \\ &= \frac{1}{2}\left(\frac{\partial u}{\partial x} - \frac{\partial v}{\partial y}\right) + \frac{i}{2}\left(\frac{\partial v}{\partial x} + \frac{\partial u}{\partial y}\right) \end{aligned} \tag{3.22}$$

Now we are in a position to determine whether or not a function is analytic—that is, if it has no dependence on $\bar{z}$—by examining how it depends on the real variables

x and y. No dependence on $\bar{z}$ implies that the real and imaginary parts of Eq. (3.22) must independently vanish. This gives us the Cauchy-Riemann equations.

Definition: The Cauchy-Riemann Equations

Using Eq. (3.22) the requirement that $\partial f/\partial \bar{z} = 0$ leads to $(\partial u/dx) - (\partial v/\partial y) = 0$ and $(\partial v/\partial x) + (\partial u/\partial y) = 0$. This gives the Cauchy-Riemann equations:

$$\begin{aligned} \frac{\partial u}{\partial x} &= \frac{\partial v}{\partial y} \\ \frac{\partial u}{\partial y} &= -\frac{\partial v}{\partial x} \end{aligned} \tag{3.23}$$

The Cauchy-Riemann equations can also be derived by looking at limits. We do this by writing everything in terms of real and imaginary parts to that once again we take $f(z) = u(x, y) + iv(x, y)$, $z_0 = x_0 + iy_0$, and $\Delta z = \Delta x + i\Delta y$. Now following what we did earlier and taking $w = f(z)$ for notational convenience, we have

$$\begin{aligned} \Delta w &= f(z_0 + \Delta z) - f(z_0) \\ &= u(x_0 + \Delta x, y_0 + \Delta y) - u(x_0, y_0) + i[v(x_0 + \Delta x, y_0 + \Delta y) - v(x_0, y_0)] \end{aligned}$$

If the function $f(z)$ is differentiable, we can approach the origin in any way we like. So we try this in two ways, first along the x axis (and hence setting $\Delta y = 0$) and then along the y axis (and setting $\Delta x = 0$). Going with the first case, we set $\Delta y = 0$ and get

$$\frac{\Delta w}{\Delta z} = \frac{u(x_0 + \Delta x, y_0) - u(x_0, y_0)}{\Delta x} + i\frac{[v(x_0 + \Delta x, y_0) - v(x_0, y_0)]}{\Delta x}$$

Now, taking the limit $\Delta x \to 0$ we see that these expressions are nothing other than partial derivatives. That is:

$$\begin{aligned} \lim_{\Delta x \to 0} \frac{\Delta w}{\Delta z} &= \lim_{\Delta x \to 0} \frac{u(x_0 + \Delta x, y_0) - u(x_0, y_0)}{\Delta x} + i \lim_{\Delta x \to 0} \frac{[v(x_0 + \Delta x, y_0) - v(x_0, y_0)]}{\Delta x} \\ &= \frac{\partial u}{\partial x} + i\frac{\partial v}{\partial x} \end{aligned}$$

If the derivative $f'(z)$ exists, we must obtain the same answer even if we choose another way for $(\Delta x, \Delta y)$ to go to zero. Now we use the other option, approaching the origin along the y axis. Hence we set $\Delta x = 0$. This time we have

$$\frac{\Delta w}{\Delta z} = \frac{u(x_0, y_0 + \Delta y) - u(x_0, y_0)}{i\Delta y} + \frac{[v(x_0, y_0 + \Delta y) - v(x_0, y_0)]}{\Delta y}$$

Taking the limit as $\Delta y \to 0$, once again we obtain partial derivatives. This time, however, they are with respect to y:

$$\lim_{\Delta y \to 0} \frac{\Delta w}{\Delta z} = \lim_{\Delta y \to 0} \frac{u(x_0, y_0 + \Delta y) - u(x_0, y_0)}{i\Delta y} + \lim_{\Delta y \to 0} \frac{[v(x_0, y_0 + \Delta y) - v(x_0, y_0)]}{\Delta y}$$
$$= -i\frac{\partial u}{\partial y} + \frac{\partial v}{\partial y}$$

This must agree with our previous result. It can only do so if the real and imaginary parts of both limits are equal. Imposing this condition on the real part of each limit we obtain the first of the Cauchy-Riemann equations:

$$\frac{\partial u}{\partial x} = \frac{\partial v}{\partial y}$$

Equating imaginary parts, we find the second of the Cauchy-Riemann equations:

$$\frac{\partial u}{\partial y} = -\frac{\partial v}{\partial x}$$

EXAMPLE 3.6

Is the function $f(z) = z^2$ analytic?

SOLUTION

Writing the function in terms of real and imaginary parts, we have

$$f(z) = z^2 = (x + iy)(x + iy)$$
$$= x^2 - y^2 + i2xy$$

Hence, in this case

$$u(x, y) = x^2 - y^2 \qquad v(x, y) = 2xy$$

Now let's compute the relevant partial derivatives. We find

$$\frac{\partial u}{\partial x} = 2x \qquad \frac{\partial u}{\partial y} = -2y$$
$$\frac{\partial v}{\partial x} = 2y \qquad \frac{\partial v}{\partial y} = 2x$$

We see immediately that

$$\frac{\partial u}{\partial x}=\frac{\partial v}{\partial y} \qquad \text{and} \qquad \frac{\partial u}{\partial y}=-\frac{\partial v}{\partial x}$$

The Cauchy-Riemann equations are satisfied, so we conclude the function is analytic.

EXAMPLE 3.7
Is $f(z)=|z|^2$ analytic?

SOLUTION
This time the situation is a little bit different. We will see that the function is only differentiable at the origin. Again, we write the function in terms of real and imaginary parts:

$$f(z)=|z|^2=z\overline{z}=(x+iy)(x-iy)=x^2+y^2$$

Therefore

$$u(x,y)=x^2+y^2 \qquad v(x,y)=0$$

So while

$$\frac{\partial u}{\partial x}=2x \qquad \text{and} \qquad \frac{\partial u}{\partial y}=2y$$

We have

$$\frac{\partial v}{\partial x}=\frac{\partial v}{\partial y}=0$$

So unfortunately, the Cauchy-Riemann equations cannot be satisfied, unless $x=y=0$. We conclude the function is not analytic. Another way to look at this is that the function has $\overline{z}$ dependence:

$$\frac{\partial f}{\partial \overline{z}}=\frac{\partial}{\partial \overline{z}}(z\overline{z})=z\neq 0$$

(unless of course, $z=0$.) This illustrates the fact that a function which depends on $\overline{z}$ is not analytic.

EXAMPLE 3.8

Is $f(z)=e^z$ analytic?

SOLUTION

We write the function in terms of real and imaginary parts as

$$e^z = e^{x+iy} = e^x e^{iy}$$

Now use Euler's formula to write

$$e^{iy} = \cos y + i \sin y$$

So we have

$$e^z = e^x(\cos y + i \sin y) = e^x \cos y + ie^x \sin y$$

Therefore

$$u(x,y) = e^x \cos y \qquad v(x,y) = e^x \sin y$$

We find the following partial derivatives of these functions:

$$\frac{\partial u}{\partial x} = e^x \cos y \qquad \frac{\partial u}{\partial y} = -e^x \sin y$$
$$\frac{\partial v}{\partial x} = e^x \sin y \qquad \frac{\partial v}{\partial y} = e^x \cos y$$

Since $u_x = v_y$ and $u_y = -v_x$, the Cauchy-Riemann equations are satisfied, and we conclude the function is analytic.

Definition: The Derivative of a Continuously Differentiable Function

If the partial derivatives u_x, u_y, v_x, and v_y are continuous at a point (x_0, y_0) and the Cauchy-Riemann equations hold then

$$\begin{aligned} f'(z_0) &= u_x(x_0,y_0) + iv_x(x_0,y_0) \\ f'(z_0) &= v_y(x_0,y_0) - iu_y(x_0,y_0) \end{aligned} \tag{3.24}$$

It follows from the Cauchy-Riemann equations that we can write

$$\frac{df}{dz} = \frac{\partial f}{\partial x} - i\frac{\partial f}{\partial y} \tag{3.25}$$

The Polar Representation

In many cases it is convenient to work with the polar representation of a complex function where we write z in the form $z = re^{i\theta}$. Then

$$f(z) = u(r,\theta) + iv(r,\theta) \tag{3.26}$$

In this case the Cauchy-Riemann equations assume the form:

$$\frac{\partial u}{\partial r} = \frac{1}{r}\frac{\partial v}{\partial \theta} \qquad \frac{\partial v}{\partial r} = -\frac{1}{r}\frac{\partial u}{\partial \theta} \tag{3.27}$$

These equations hold provided that $f(z)$ is defined throughout an ε neighborhood of a nonzero point $z_0 = r_0 e^{i\theta_0}$ and the first-order partial derivatives u_r, v_r, u_θ, and v_θ exist and are continuous everywhere in the ε neighborhood.

EXAMPLE 3.9

Let f be the principal square root function

$$f(z) = \sqrt{z}$$

with $z = re^{i\theta}$ defined such that $r > 0$ and $-\pi < \theta < \pi$. Is this function analytic?

SOLUTION

We write the function in terms of the polar representation:

$$f(z) = \sqrt{z} = \sqrt{re^{i\theta}} = \sqrt{r}\, e^{i\theta/2}$$

Using Euler's formula this can be written as

$$f(z) = \sqrt{r}\, e^{i\theta/2} = \sqrt{r}\cos(\theta/2) + i\sqrt{r}\sin(\theta/2)$$

So we have

$$u(r,\theta) = \sqrt{r}\cos(\theta/2) \qquad \text{and} \qquad v(r,\theta) = \sqrt{r}\sin(\theta/2)$$

This means that

$$\frac{\partial u}{\partial r} = \frac{1}{2\sqrt{r}}\cos(\theta/2) \qquad \frac{\partial u}{\partial \theta} = -\frac{\sqrt{r}}{2}\sin(\theta/2)$$
$$\frac{\partial v}{\partial r} = \frac{1}{2\sqrt{r}}\sin(\theta/2) \qquad \frac{\partial v}{\partial \theta} = \frac{\sqrt{r}}{2}\cos(\theta/2)$$

Since we have the following relationships:

$$ru_r = \frac{\sqrt{r}}{2}\cos(\theta/2) = v_\theta$$
$$u_\theta = -\frac{\sqrt{r}}{2}\sin(\theta/2) = -rv_r$$

The polar form of the Cauchy-Riemann equations are satisfied and the given function is analytic on the specified domain.

In an analogous manner to Eq. (3.25), using the Cauchy-Riemann equations it can be shown that

$$f'(z) = e^{-i\theta}\left(\frac{\partial u}{\partial r} + i\frac{\partial v}{\partial r}\right) \tag{3.28}$$

EXAMPLE 3.10

Does the derivative of $f(z) = 1/z$ exist? If so what is $f'(z)$ in polar form?

SOLUTION

First let's write the function in polar form:

$$f(z) = \frac{1}{z} = \frac{1}{re^{i\theta}} = \frac{1}{r}e^{-i\theta}$$

Using Euler's formula, we can split this into real and imaginary parts:

$$f(z) = \frac{1}{r}\cos\theta - i\frac{1}{r}\sin\theta$$

Computing the derivatives we find

$$\frac{\partial u}{\partial r} = -\frac{1}{r^2}\cos\theta \qquad \frac{\partial u}{\partial \theta} = -\frac{1}{r}\sin\theta$$
$$\frac{\partial v}{\partial r} = \frac{1}{r^2}\sin\theta \qquad \frac{\partial v}{\partial \theta} = -\frac{1}{r}\cos\theta$$

It's apparent that these results satisfy

$$r\frac{\partial u}{\partial r} = \frac{\partial v}{\partial \theta} \quad \text{and} \quad -r\frac{\partial v}{\partial r} = \frac{\partial u}{\partial \theta}$$

This means that the derivative exists, since the Cauchy-Riemann equations are satisfied (provided that $r > 0$). Using Eq. (3.28) we can find $f'(z)$. We obtain

$$\begin{aligned}
f'(z) &= e^{-i\theta}\left(\frac{\partial u}{\partial r} + i\frac{\partial v}{\partial r}\right) = e^{-i\theta}\left(-\frac{1}{r^2}\cos\theta + \frac{i}{r^2}\sin\theta\right) \\
&= -\frac{e^{-i\theta}}{r^2}(\cos\theta - i\sin\theta) \\
&= -\frac{e^{-i\theta}}{r^2}(e^{-i\theta}) \\
&= -\frac{e^{-i2\theta}}{r^2} \\
&= -\frac{1}{r^2 e^{i2\theta}} = -\frac{1}{z^2}
\end{aligned}$$

Some Consequences of the Cauchy-Riemann Equations

Let's take a step back and formally define the term analytic. We say that a function $f(z)$ is analytic in an open set S if its derivative exists and is continuous at every point in S. If you hear a mathematician say that a complex function is regular or *holomorphic,* the meaning is the same.

A function can't be analytic if its derivative only exists at a point. If we say that $f(z)$ is analytic at some point z_0 then this means it is analytic throughout some neighborhood of z_0. So, recalling Example 3.7, while $f(z) = |z|^2$ satisfies the Cauchy-Riemann equations at the origin, it is not analytic because it does not satisfy the Cauchy-Riemann equations at any point displaced from the origin (or at any nonzero point). As a result we cannot construct a neighborhood about the origin where the Cauchy-Riemann equations would be satisfied, so the function is not analytic.

Definition: Singularity

Suppose that a function $f(z)$ is not analytic at some point z_0, but it's analytic in a neighborhood that contains z_0. In this case, we say that z_0 is a *singularity* or *singular point* of $f(z)$. Singularities will take center stage when we talk about power series expansions of complex functions.

Definition: Necessary and Sufficient Conditions for a Function to Be Analytic

There are two *necessary* conditions a function $f(z)$ must satisfy to be analytic. These are

- $f(z)$ must be continuous
- The Cauchy-Riemann equations must be satisfied

These conditions, however, are not *sufficient* to say a function is analytic. To satisfy the sufficiency condition for differentiability at a point z_0, a function $f(z)$ must satisfy the following conditions:

- It must be defined throughout an ε-neighborhood of the point z_0.
- The first-order partial derivatives u_x, u_y, v_x, and v_y must exist everywhere throughout the ε-neighborhood.
- The partial derivatives must be continuous at z_0 and the Cauchy-Riemann equations must be satisfied.

SOME PROPERTIES OF ANALYTIC FUNCTIONS

Let $f(z)$ and $g(z)$ be two analytic functions on some domain *D*. Then

- The sum and difference $f \pm g$ is also analytic in *D*.
- The product $f(z)g(z)$ is analytic in *D*.
- If $g(z)$ does not vanish at any point in *D*, then the quotient $f(z)/g(z)$ is analytic in *D*.
- The composition of two analytic functions $g[f(z)]$ or $f[g(z)]$ is analytic in *D*

EXAMPLE 3.11

Determine whether or not the function $f(z) = (z^2+1)/[(z+2)(z^2+3)]$ is analytic.

SOLUTION

Since z^2+1 and $(z+2)(z^2+3)$ are both analytic (note there is no $\bar{z}$ dependence), and the quotient of two analytic functions is analytic, we conclude that $f(z)$ is analytic. However, this is not true at any singular points of the function, which are points for which the denominator vanishes. The singular points in this case are

$$z = -2 \qquad \pm i\sqrt{3}$$

So, we say that $f(z)$ is analytic throughout the complex plane except at these points, which are the singularities of the function.

An important theorem which is a consequence of the Cauchy-Riemann equations tells us if a function is constant in a domain D.

THEOREM 3.3

If $f'(z) = 0$ everywhere in a domain D, then $f(z)$ must be constant in D.

Harmonic Functions

An important class of functions known as *harmonic functions* play an important role in many areas of applied mathematics, physics, and engineering. We say that a function $u(x, y)$ is a harmonic function if it satisfies Laplace's equation in some domain of the x-y plane:

$$\frac{\partial^2 u}{\partial x^2} + \frac{\partial^2 u}{\partial y^2} = 0 \tag{3.29}$$

Here, we have assumed that $u(x, y)$ has continuous first and second partial derivatives with respect to both x and y. An important application of harmonic functions in physics and engineering is in the area of electrostatics, for example. Harmonic functions also find application in the study of temperature and fluid flow.

It turns out that the Cauchy-Riemann equations can help us find harmonic functions, as the next theorem illustrates.

THEOREM 3.4

Suppose that $f(z) = u(x, y) + iv(x, y)$ is an analytic function in a domain D. If follows that $u(x, y)$ and $v(x, y)$ are harmonic functions.

PROOF

The proof is actually easy. Since $f(z)$ is analytic, then the Cauchy-Riemann equations are satisfied:

$$\frac{\partial u}{\partial x} = \frac{\partial v}{\partial y} \quad \text{and} \quad \frac{\partial u}{\partial y} = -\frac{\partial v}{\partial x}$$

Now let's take the derivative of the first equation with respect to x:

$$\frac{\partial^2 u}{\partial x^2} = \frac{\partial^2 v}{\partial x \partial y}$$

Taking the derivative of the second Cauchy-Riemann equation with respect to y gives

$$\frac{\partial^2 u}{\partial y^2} = -\frac{\partial^2 v}{\partial y \partial x}$$

Since partial derivatives commute (that is, their order does not matter) it is the case that

$$\frac{\partial^2 v}{\partial x \partial y} = \frac{\partial^2 v}{\partial y \partial x}$$

From which it follows that

$$\frac{\partial^2 u}{\partial x^2} = -\frac{\partial^2 u}{\partial y^2}$$

$$\Rightarrow \frac{\partial^2 u}{\partial x^2} + \frac{\partial^2 u}{\partial y^2} = 0$$

So we've shown that if a function is analytic, then the real part $u(x, y)$ satisfies Laplace's equation. A similar procedure can be used to show that the imaginary part $v(x, y)$ is harmonic as well.

Definition: Harmonic Conjugate

Suppose that u and v are two harmonic functions in a domain D. If their first-order partial derivatives satisfy the Cauchy-Riemann equations, then we say that v is the harmonic conjugate of u.

THEOREM 3.5

A function $f(z) = u(x, y) + iv(x, y)$ is analytic if and only if $v(x, y)$ is the harmonic conjugate of $u(x, y)$.

EXAMPLE 3.12

Is $u(x, y) = e^{-y} \sin x$ a harmonic function? If so write down an analytic function $f(z)$ such that u is its real part.

SOLUTION

We compute the partial derivatives of u. We find

$$\frac{\partial u}{\partial x} = e^{-y} \cos x \qquad \Rightarrow \frac{\partial^2 u}{\partial x^2} = -e^{-y} \sin x$$

$$\frac{\partial u}{\partial y} = -e^{-y} \sin x \qquad \Rightarrow \frac{\partial^2 u}{\partial y^2} = e^{-y} \sin x$$

It's clear that the function is harmonic since

$$\frac{\partial^2 u}{\partial x^2} + \frac{\partial^2 u}{\partial y^2} = -e^{-y}\sin x + e^{-y}\sin x = 0$$

Using the Cauchy-Riemann equations, we have

$$\frac{\partial u}{\partial x} = e^{-y}\cos x = \frac{\partial v}{\partial y}$$
$$\frac{\partial u}{\partial y} = -e^{-y}\sin x = -\frac{\partial v}{\partial x}$$

So it must be the case that $v(x, y) = -e^{-y}\cos x$. The analytic function we seek is therefore

$$f(z) = u(x, y) + iv(x, y) = e^{-y}\sin x - ie^{-y}\cos x$$

The Reflection Principle

The *reflection principle* allows us to determine when the following condition is satisfied:

$$\overline{f(z)} = f(\overline{z}) \tag{3.30}$$

If $f(z)$ is analytic in a domain D that contains a segment of the x axis, then Eq. (3.30) holds if and only if $f(x)$ is real for each point of the segment of x contained in D.

EXAMPLE 3.13

Do $f(z) = z + 1$ and $g(z) = z + i$ satisfy the reflection principle?

SOLUTION

Since $f(x) = x + 1$ is a real number in all cases, the reflection principle is satisfied. In this simple example we can actually see this immediately since $\overline{f(z)} = \overline{z+1} = \overline{z} + 1 = f(\overline{z})$. In the second case, we have $g(x) = x + i$, which is not a real number. The reflection principle is not satisfied which means that $\overline{g(z)} \neq g(\overline{z})$. This function is also simple enough so that we can verify this explicitly—we have $\overline{g(z)} = \overline{z+i} = \overline{z} - i$, but $g(\overline{z}) = \overline{z} + i$.

Summary

In this chapter, we learned how to determine if a function of a complex variable is differentiable or analytic. The necessary condition for a function to be analytic is that it be continuous and satisfy the Cauchy-Riemann equations. If a function is analytic, then its real and imaginary parts are harmonic functions, that is, they satisfy Laplace's equation.

Quiz

1. Let $f(z) = z^n$. Using the limit procedure outlined in Example 3.1, find $f'(z)$.
2. Let $f(z) = |z|^2$. Find $\dfrac{\Delta w}{\Delta z}$. Does the derivative exist?
3. Compute $f'(z)$ when $f(z) = 3z^8 - 6z^2 + 4$.
4. What is the derivative of $f(z) = \dfrac{(3+2z)^2}{z^2}$?
5. What is $\lim_{z\to 1} \dfrac{z-1}{z^2 + iz - i + 1}$?
6. Show that $e^{\bar{z}}$ is not analytic using the Cauchy-Riemann equations.
7. Is $f = x - iy$ differentiable at the origin?
8. Let $f(z) = e^{iz}$. Is this function entire?
9. Does $f = \sqrt[3]{r}e^{i\theta/3}$ have a derivative everywhere in the domain $r > 0, 0 < \theta < 2\pi$?
10. Let $u(x, y) = x^2 - y^2$. Is this function harmonic? If so, what is the harmonic conjugate?

CHAPTER 4

Elementary Functions

In this chapter we introduce some of the elementary functions in the context of complex analysis. Our discussion will include polynomials, rational functions, the exponential and logarithm, trigonometric functions and their inverses, and finally, the hyperbolic functions and their inverses.

Complex Polynomials

A *polynomial* is a function $f(z)$ that can be written in the form

$$f(z) = a_0 + a_1 z + a_2 z^2 + \cdots + a_n z^n \tag{4.1}$$

The highest power n is called the *degree* of the polynomial and a_j are constants called *coefficients*. In general, the coefficients can be complex numbers.

Since $z = x + iy$, a complex polynomial can be viewed as a polynomial in the real variables x and y with complex coefficients. For example, consider

$$f(z) = 5 - iz + 2z^2 \tag{4.2}$$

We can write this as

$$\begin{aligned} f(z) &= 5 - iz + 2z^2 \\ &= 5 - i(x + iy) + 2(x + iy)^2 \\ &= 5 - ix + y + 2x^2 + i4xy - 2y^2 \\ &= 5 + 2x^2 - 2y^2 + y + i(4xy - x) \end{aligned}$$

Following the last chapter, we can identify

$$\begin{aligned} & u(x, y) = 5 + 2x^2 - 2y^2 + y \\ & \Rightarrow \frac{\partial u}{\partial x} = 4x \qquad \frac{\partial u}{\partial y} = -4y + 1 \end{aligned}$$

and

$$\begin{aligned} & v(x, y) = 4xy - x \\ & \Rightarrow \frac{\partial v}{\partial x} = 4y - 1 \qquad \frac{\partial v}{\partial y} = 4x \end{aligned}$$

Notice that $u_x = v_y$ and $u_y = -v_x$, so this is an analytic function. We can also verify this by noticing that

$$\frac{\partial f}{\partial \bar{z}} = \frac{\partial}{\partial \bar{z}}(5 - iz + 2z^2) = 0$$

We can generate some plots to look at the behavior of the function. When studying a complex function, you might want to plot its modulus and the real and imaginary parts to see where interesting features appear. Let's plot $|f(z)| = \sqrt{f(z)\overline{f(z)}}$. This is shown in Fig. 4.1.

Obviously, the function has some interesting behavior around the origin. We see more by looking at the contour plot, shown in Fig. 4.2.

The most interesting behavior seems to be around $x = 0$. At this point, the real and imaginary parts of the function are given by

$$u(0, y) = 5 - 2y^2 + y \qquad v(0, y) = 0$$

Plotting the modulus of f with $x = 0$, we see the two zeros in y as shown in Fig. 4.3.

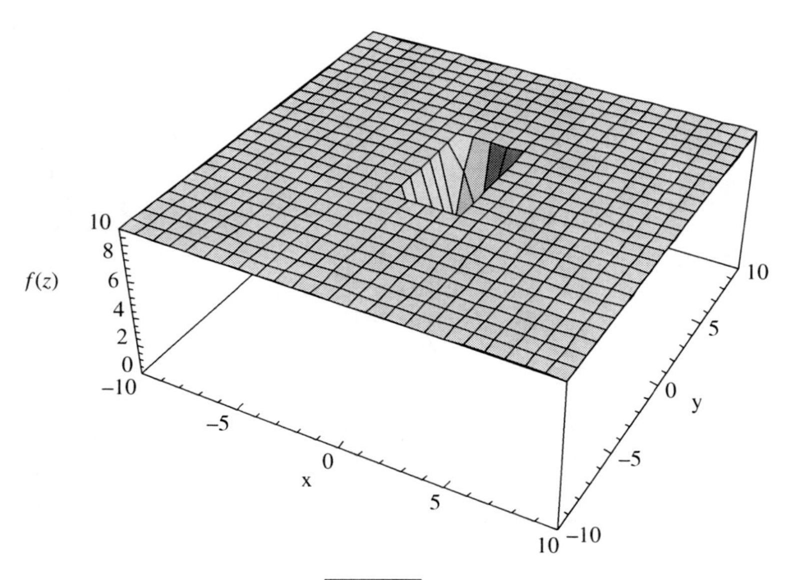

Figure 4.1 A plot of $|f(z)| = \sqrt{f(z)\overline{f(z)}}$ for the function defined in Eq. (4.2).

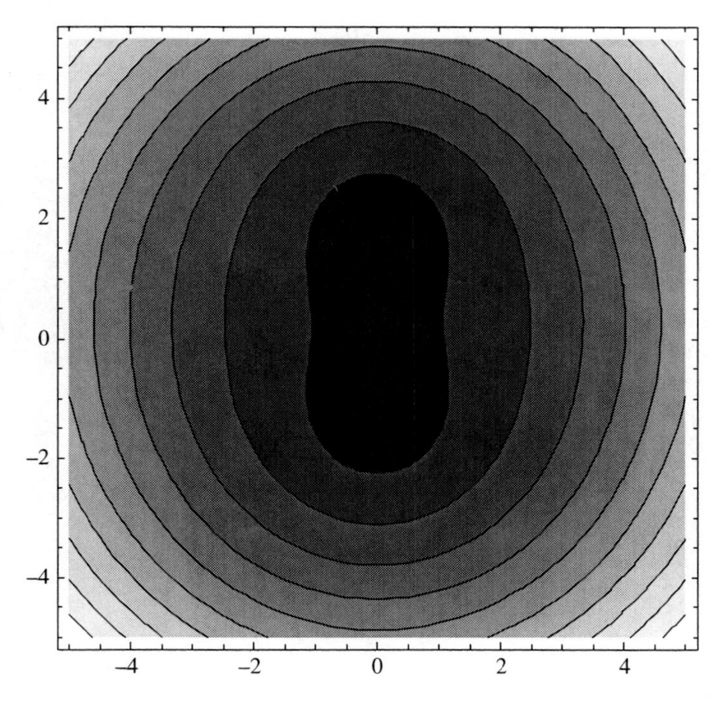

Figure 4.2 A contour plot showing the modulus of $f(z) = 5 - iz + 2z^2$.

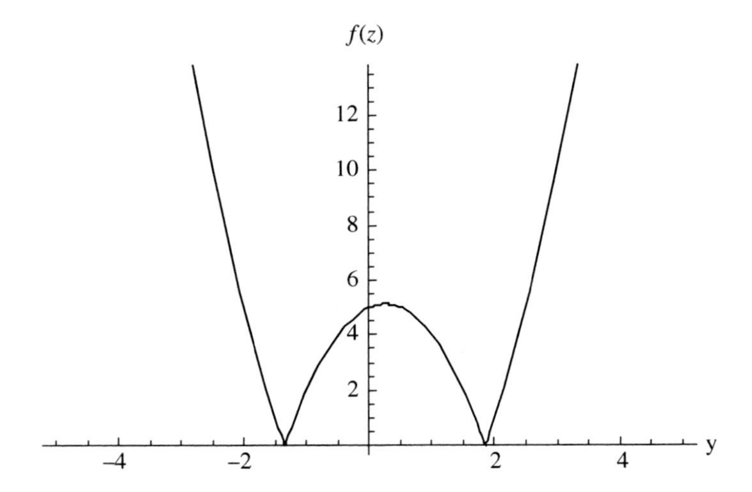

Figure 4.3 A plot of the modulus of $f(z)$ with $x = 0$.

You can also look at the real and imaginary parts to learn about the function. In Fig. 4.4, we show a plot of the real part of Eq. (4.2) and in Fig. 4.5, we show a contour plot of the real part of the function. A plot of the imaginary part of Eq. (4.2) and its contours are shown in Figs. 4.6 and 4.7, respectively.

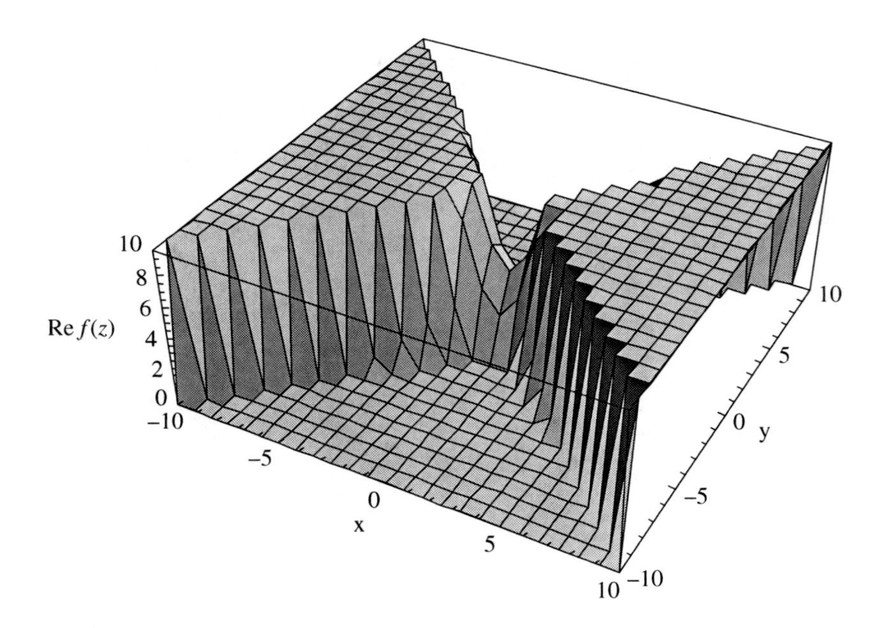

Figure 4.4 The real part of $f(z)$.

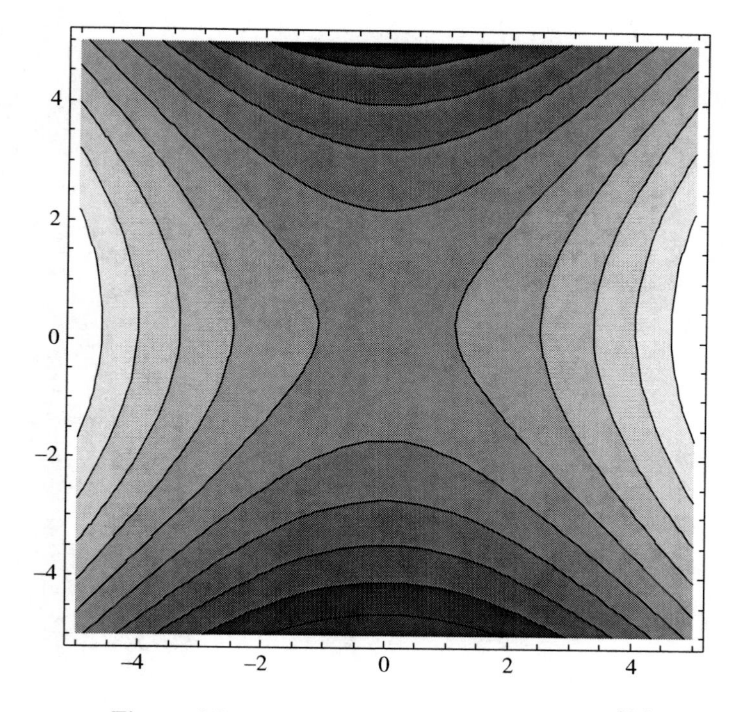

Figure 4.5 A contour plot of the real part of $f(z)$.

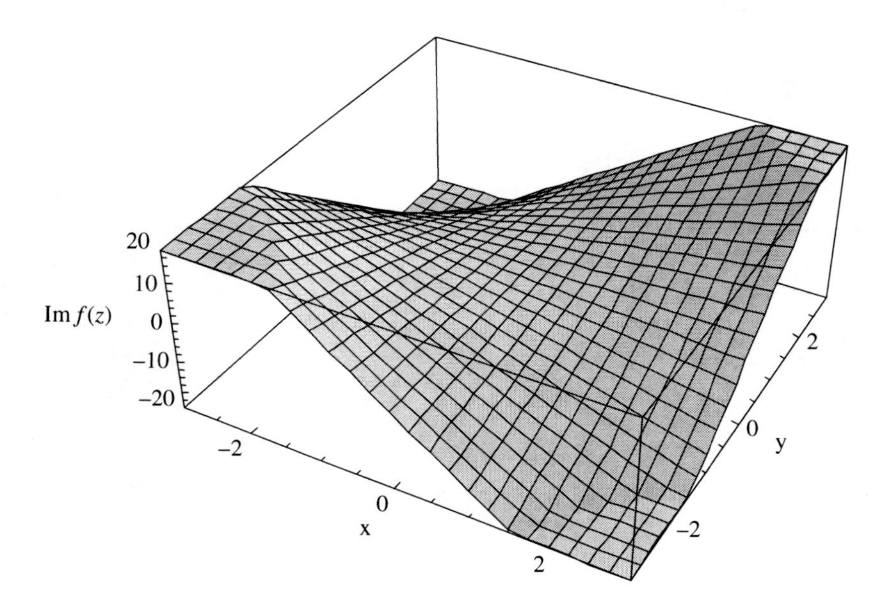

Figure 4.6 The imaginary part of Eq. (4.2).

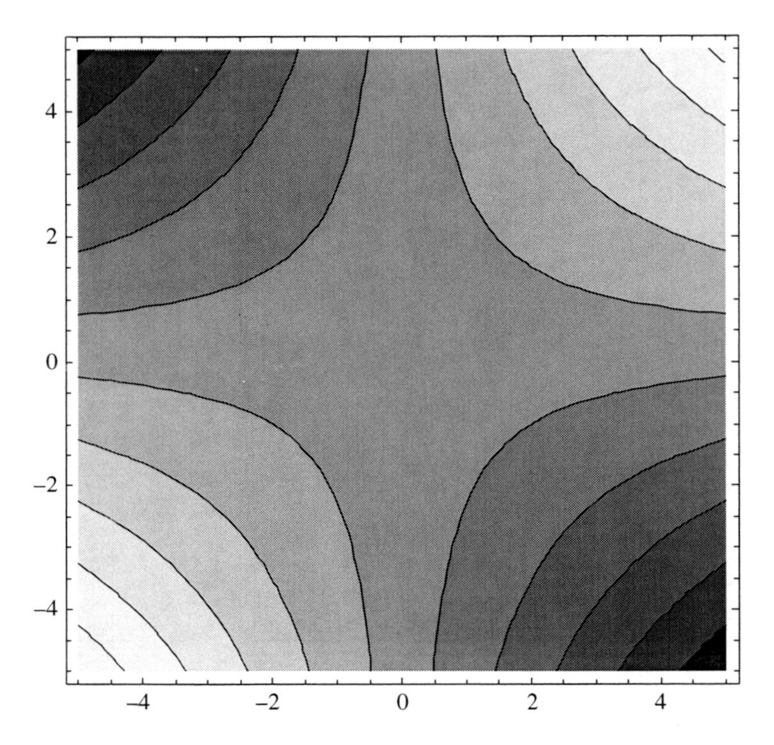

Figure 4.7 A contour plot of the imaginary part of the function.

The Complex Exponential

We have already seen the exponential function e^z. In this section, we review some of it's properties. Using $z = x + iy$ we have already noted that the complex exponential can be written as

$$e^z = e^x e^{iy} \tag{4.3}$$

Using Euler's formula we have

$$e^{iy} = \cos y + i \sin y \tag{4.4}$$

Expanding out Eq. (4.3) we have the real and imaginary parts of the complex exponential:

$$\begin{aligned} \mathrm{Re}(e^z) &= u(x, y) = e^x \cos y \\ \mathrm{Im}(e^z) &= v(x, y) = e^x \sin y \end{aligned} \tag{4.5}$$

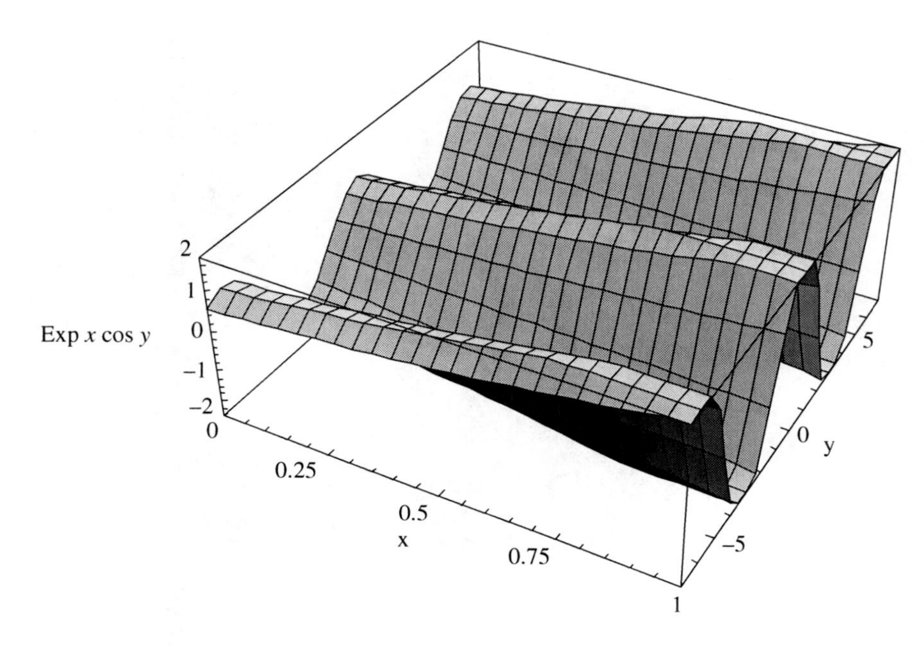

Figure 4.8 A plot of the real part of e^z.

Looking at these functions, you can see that both the real and imaginary parts increase without bound as $x \to \infty$. In Fig. 4.8, we show a plot of the real part of e^z for $0 \le x \le 1, -2\pi \le y \le 2\pi$. The oscillations due to the cos function in the y direction are readily apparent, as is the fact that the function is increasing rapidly in the x direction.

A contour plot of the real part of e^z is shown in Fig. 4.9. The oscillations along the y direction are apparent.

The additive property of exponents, that is, $e^a e^b = e^{a+b}$ also carries over to the complex case. This is due to the way we add complex numbers. Let $z_1 = x_1 + iy_1$ and $z_2 = x_2 + iy_2$. Then we know that $z_1 + z_2 = (x_1 + x_2) + i(y_1 + y_2)$. Now we utilize the fact that exponents add for real numbers. That is:

$$\begin{aligned} e^{z_1} e^{z_2} &= e^{x_1 + iy_1} e^{x_2 + iy_2} \\ &= (e^{x_1} e^{iy_1})(e^{x_2} e^{iy_2}) \\ &= (e^{x_1} e^{x_2})(e^{iy_1} e^{iy_2}) \\ &= e^{x_1 + x_2} (e^{iy_1} e^{iy_2}) \end{aligned}$$

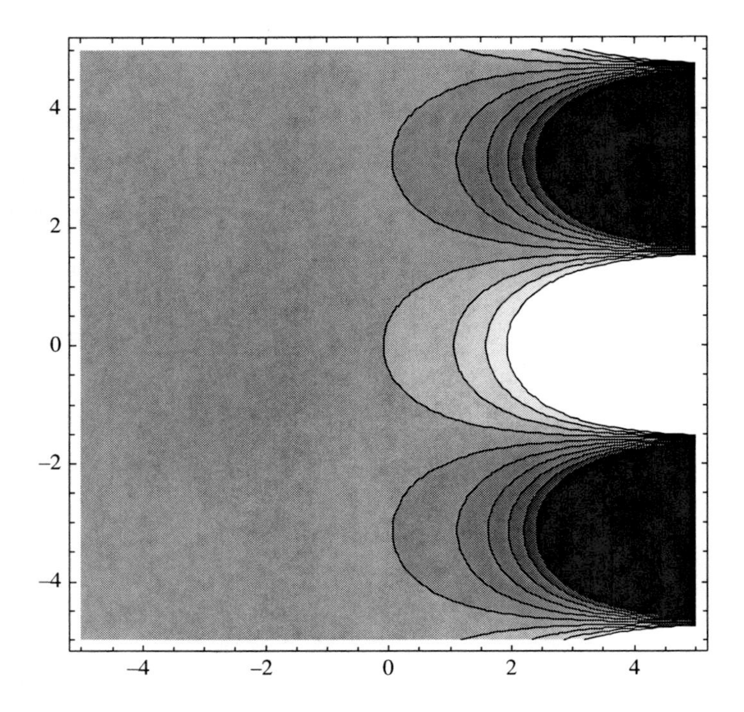

Figure 4.9 A contour plot of the real part of e^z.

We can't just assume the additive property holds for $e^{iy_1}e^{iy_2}$. But we can show it does fairly easily:

$$\begin{aligned} e^{iy_1}e^{iy_2} &= (\cos y_1 + i\sin y_1)(\cos y_2 + i\sin y_2) \\ &= \cos y_1 \cos y_2 - \sin y_1 \sin y_2 + i(\sin y_1 \cos y_2 + \cos y_1 \sin y_2) \\ &= \cos(y_1 + y_2) + i\sin(y_1 + y_2) \\ &= \exp[i(y_1 + y_2)] \end{aligned}$$

So we've got

$$\begin{aligned} e^{z_1}e^{z_2} &= e^{x_1+x_2}(e^{iy_1}e^{iy_2}) \\ &= e^{x_1+x_2}e^{i(y_1+y_2)} = e^{z_1+z_2} \end{aligned}$$

It follows that

$$\frac{e^{z_1}}{e^{z_2}} = e^{z_1-z_2} \tag{4.6}$$

Using $e^0 = 1$, you can deduce from Eq. (4.6) that $1/e^z = e^{-z}$ as in the real case.

EXAMPLE 4.1

Evaluate $\exp[(1 + \pi i)/4]$.

SOLUTION

We write this as

$$\exp\left(\frac{1+\pi i}{4}\right) = e^{1/4} e^{i\pi/4} = \sqrt[4]{\frac{e}{4}}\,(e^{i\pi/4})$$
$$= \sqrt[4]{e}\,(\cos \pi/4 + i \sin \pi/4)$$
$$= \sqrt[4]{e}\left(\frac{1}{\sqrt{2}} + i\frac{1}{\sqrt{2}}\right) = \sqrt[4]{e}\,\frac{1}{\sqrt{2}}(1+i)$$

EXAMPLE 4.2

Find $|e^z|$.

SOLUTION

We have

$$e^z = e^{x+iy} = e^x e^{iy} = e^x(\cos y + i \sin y)$$

So

$$|e^z| = |e^x(\cos y + i \sin y)|$$
$$= |e^x||\cos y + i \sin y|$$
$$= e^x \cdot 1$$
$$= e^x$$

EXAMPLE 4.3

Show that e^z is a periodic function with period $2k\pi i$, where k is an integer.

SOLUTION

We have

$$e^{z+2k\pi i} = e^z e^{2k\pi i}$$
$$= e^z(\cos 2k\pi + i \sin 2k\pi)$$
$$= e^z$$

EXAMPLE 4.4

What is the argument of e^z?

SOLUTION

In polar coordinates, a complex variable is written as

$$z = re^{i\theta}$$

Hence

$$e^z = |e^z| e^{i\theta}$$

for some θ which is $\arg(e^z)$. We have already seen in Example 4.2 that $|e^z| = e^x$. We also know that

$$e^z = e^x e^{iy}$$

Therefore, comparing with $e^z = |e^z| e^{i\theta}$, we conclude that $\arg(e^z) = y$. But we aren't quite done. Since the cosine and sin functions are 2π periodic, and $e^{iy} = \cos y + i \sin y$, we can add any integer multiple of 2π to the argument without changing anything. So the argument is really given by

$$\arg(e^z) = y + 2n\pi \qquad n = 0, \pm 1, \pm 2, \ldots$$

You know from elementary calculus that the *logarithm* is the inverse, if you will, of the exponential. That is:

$$e^{\ln x} = x$$

A similar function exists in complex variables. Due to the periodic nature of e^z, we will see that the complex logarithm is a multivalued function. We define the natural logarithm in the following way. Let $z = e^w$. Then

$$w = \ln z \tag{4.7}$$

Now use the polar representation of z, namely, $z = re^{i\theta}$. Now we have

$$w = \ln z = \ln(re^{i\theta}) = \ln r + \ln e^{i\theta} = \ln r + i\theta \tag{4.8}$$

Using the fact that the cosine and sin functions are 2π periodic, the correct representation is actually

$$w = \ln r + i(\theta + 2k\pi) \qquad k = 0, \pm 1, \pm 2, \ldots \tag{4.9}$$

A key aspect of definition in Eq. (4.9) is that the complex natural logarithm is a multivalued function. Definition in Eq. (4.8), for which $k = 0$, is called the *principal value* or the *principal branch* of $\ln z$. In that case, we are restricting the argument to $0 \le \theta < 2\pi$. Note that this choice is not unique, all that is required is that we select an interval of length 2π. So it is equally valid to choose the principal branch for $-\pi < \theta \le \pi$.

Trigonometric Functions

You have already seen the use of trigonometric functions in the theory of complex variables. Here we state some familiar results for reference. First, we write the cosine and sin functions in terms of the complex exponential. This follows from Euler's identity. You should already be familiar with these results:

$$\cos z = \frac{e^{iz} + e^{-iz}}{2} \tag{4.10}$$

$$\sin z = \frac{e^{iz} - e^{-iz}}{2i} \tag{4.11}$$

The tangent function can be written in terms of exponentials using Eqs. (4.10) and (4.11):

$$\tan z = \frac{\sin z}{\cos z} = -i\frac{e^{iz} - e^{-iz}}{e^{iz} + e^{-iz}} \tag{4.12}$$

Likewise, we have the cotangent function which is just the reciprocal of Eq (4.12):

$$\cot z = \frac{\cos z}{\sin z} = i\frac{e^{iz} + e^{-iz}}{e^{iz} - e^{-iz}} \tag{4.13}$$

The secant and cosecant functions can also be written down in terms of exponentials. These are given by

$$\sec z = \frac{1}{\cos z} = \frac{2}{e^{iz} + e^{-iz}} \tag{4.14}$$

$$\csc z = \frac{1}{\sin z} = \frac{2i}{e^{iz} - e^{-iz}} \tag{4.15}$$

All of the results from trigonometry using real variables carry over to complex variables. We illustrate this in the next two examples.

EXAMPLE 4.5
Show that

$$\sin(x + iy) = \sin x \cos iy + \cos x \sin iy$$

SOLUTION
Using Euler's identity:

$$\begin{aligned}
\sin(x + iy) &= \frac{e^{i(x+iy)} - e^{-i(x+iy)}}{2i} \\
&= \frac{e^{ix}e^{-y} - e^{-ix}e^{y}}{2i} \\
&= \frac{(\cos x + i\sin x)e^{-y} - (\cos x - i\sin x)e^{y}}{2i} \\
&= \frac{i\sin x(e^{y} + e^{-y})}{2i} + \frac{\cos x(e^{-y} - e^{y})}{2i} \\
&= \sin x\frac{e^{i(iy)} + e^{-i(iy)}}{2} + \cos x\frac{e^{i(iy)} - e^{-i(iy)}}{2i} \\
&= \sin x \cos iy + \cos x \sin iy
\end{aligned}$$

EXAMPLE 4.6

Show that $\cos^2 z + \sin^2 z = 1$.

SOLUTION

We start by writing $z = x + iy$ and utilize the fact that $\cos^2 x + \sin^2 x = 1$, when x is a real variable. Using the result of the last example we have

$$\sin(x+iy) = \sin x \cos iy + \cos x \sin iy$$
$$\cos(x+iy) = \cos x \cos iy - \sin x \sin iy$$

Therefore

$$\begin{aligned}\sin^2 z &= (\sin x \cos iy + \cos x \sin iy)^2 = \sin^2 x \cos^2 iy + \cos^2 x \sin^2 iy \\ &\quad +2\cos x \sin x \cos iy \sin iy \\ \cos^2 z &= (\cos x \cos iy - \sin x \sin iy)^2 = \cos^2 x \cos^2 iy + \sin^2 x \sin^2 iy \\ &\quad -2\cos x \sin x \cos iy \sin iy\end{aligned}$$

So it follows that

$$\begin{aligned}\cos^2 z + \sin^2 z &= \sin^2 x \cos^2 iy + \cos^2 x \sin^2 iy + 2\cos x \sin x \cos iy \sin iy \\ &\quad + \cos^2 x \cos^2 iy + \sin^2 x \sin^2 iy - 2\cos x \sin x \cos iy \sin iy \\ &= \cos^2 x \cos^2 iy + \sin^2 x \cos^2 iy + \cos^2 x \sin^2 iy + \sin^2 x \sin^2 iy \\ &= \cos^2 iy(\cos^2 x + \sin^2 x) + \sin^2 iy(\cos^2 x + \sin^2 x) \\ &= \cos^2 iy + \sin^2 iy\end{aligned}$$

Now we expand each terms using Euler's identity:

$$\begin{aligned}\cos^2 iy &= \left(\frac{e^{i(iy)} + e^{-i(iy)}}{2}\right)^2 \\ &= \left(\frac{e^{-y} + e^{y}}{2}\right)^2 = \frac{e^{2y} + e^{-2y} + 2}{4}\end{aligned}$$

and

$$\begin{aligned}\sin^2 iy &= \left(\frac{e^{i(iy)} - e^{-i(iy)}}{2i}\right)^2 \\ &= \left(\frac{e^{-y} - e^{y}}{2i}\right)^2 = -\left(\frac{e^{2y} + e^{-2y} - 2}{4}\right)\end{aligned}$$

Therefore

$$\cos^2 iy + \sin^2 iy = \frac{e^{2y} + e^{-2y} + 2}{4} - \left(\frac{e^{2y} + e^{-2y} - 2}{4}\right) = \frac{1}{2} + \frac{1}{2} = 1$$

Hence, $\cos^2 z + \sin^2 z = 1$.

Following real variables, the trigonometric functions of a complex variable have inverses. Let $z = \cos w$. Then we define the inverse $w = \cos^{-1} z$, which we call the *arc cosine* function or *cosine inverse.* There is an inverse trigonometric function for each of the trigonometric functions defined in Eqs. (4.10)–(4.15). The inverses are written in terms of the complex logarithm (see Example 1.8 for a derivation). The formulas are

$$\cos^{-1} z = \frac{1}{i} \ln\left(z + \sqrt{z^2 - 1}\right) \tag{4.16}$$

$$\sin^{-1} z = \frac{1}{i} \ln\left(iz + \sqrt{1 - z^2}\right) \tag{4.17}$$

$$\tan^{-1} z = \frac{1}{2i} \ln\left(\frac{1 + iz}{1 - iz}\right) \tag{4.18}$$

$$\sec^{-1} z = \frac{1}{i} \ln\left(\frac{1 + \sqrt{1 - z^2}}{z}\right) \tag{4.19}$$

$$\csc^{-1} z = \frac{1}{i} \ln\left(\frac{i + \sqrt{z^2 - 1}}{z}\right) \tag{4.20}$$

$$\cot^{-1} z = \frac{1}{2i} \ln\left(\frac{z + i}{z - i}\right) \tag{4.21}$$

The Hyperbolic Functions

The complex hyperbolic functions are defined in terms of the complex exponential as follows:

$$\cosh z = \frac{e^z + e^{-z}}{2} \tag{4.22}$$

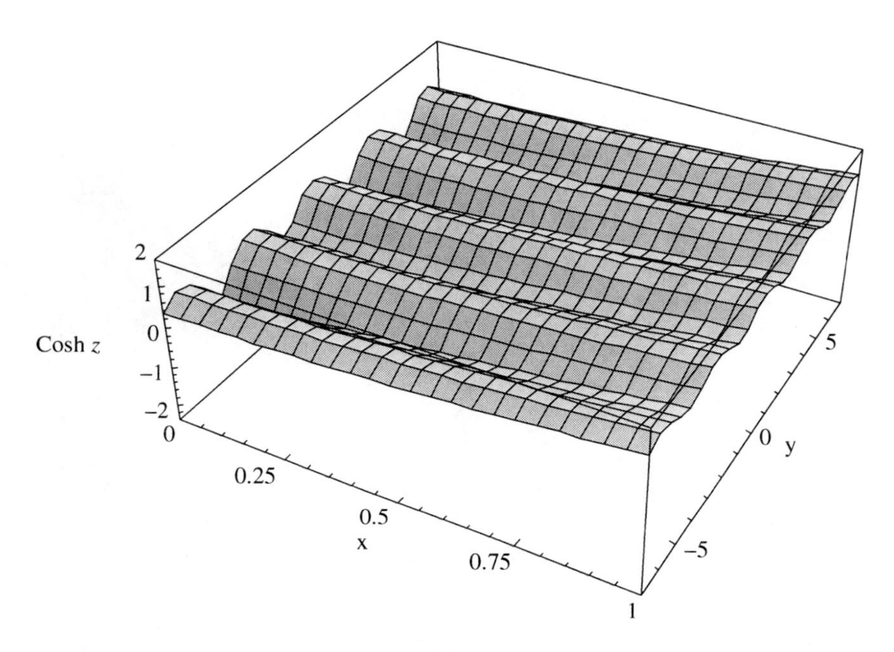

Figure 4.10 A plot of $|\cosh z|$ with $0 \le x \le 1$.

$$\sinh z = \frac{e^z - e^{-z}}{2} \tag{4.23}$$

These functions show some interesting features. Let's take a closer look at $\cosh z$. A plot of $|\cosh z|$ is shown in Fig. 4.10 focusing on the region $0 \le x \le 1$. Note the oscillations along the y direction.

These oscillations result from the fact that this function has trigonometric functions with y argument. To see this, we write the hyperbolic cosine function in terms of $z = x + iy$:

$$\begin{aligned}
\cosh z &= \frac{e^z + e^{-z}}{2} \\
&= \frac{e^{x+iy} + e^{-(x+iy)}}{2} \\
&= \frac{e^x(\cos y + i \sin y) + e^{-x}(\cos y - i \sin y)}{2} \\
&= \cos y \cosh x + i \sin y \sinh x
\end{aligned}$$

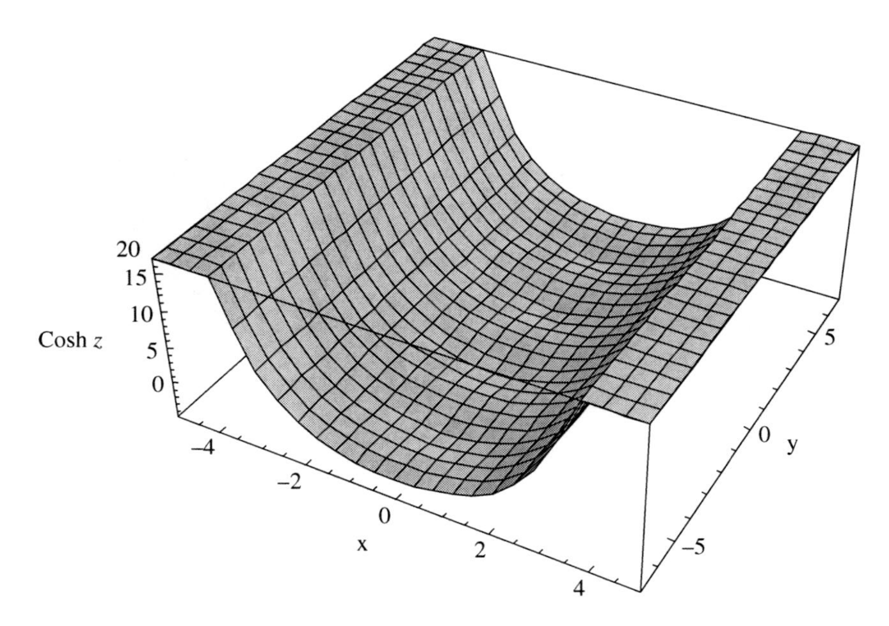

Figure 4.11 Looking at $|\cosh z|$ over a wider region, we see that the oscillations sit in a region which is surrounded by exponential growth on both sides.

From here, we see that the modulus is given by

$$|\cosh z| = \cos^2 y \cosh^2 x + \sin^2 y \sinh^2 x$$

These oscillations actually sit inside a kind of half-pipe. This is shown in Fig. 4.11.

To see what's happening, consider a plot of $\cosh^2 x + \sinh^2 x$. The function quickly grows out of control, as shown in Fig. 4.12.

If we look at the real part of $\cosh z$ alone, the oscillations are stronger. Compare Fig. 4.13, which shows the real part of the function, to Fig. 4.11, which shows the modulus over the same region. The differences are also apparent in the contour plots, which are shown side by side in Fig. 4.14. The oscillations are highly visible in the contour plot of the real part of the function, shown on the right.

The reason that the oscillations appear more prominent in plots of the real part of the function is that we have

$$\cosh z = \cos y \cosh x + i \sin y \sinh x$$

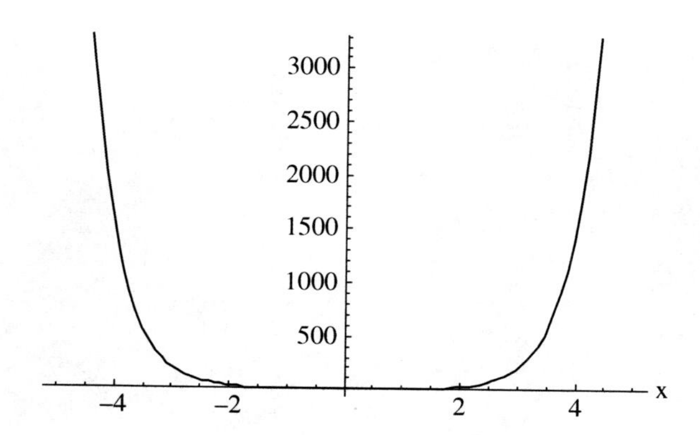

Figure 4.12 A plot of $\cosh^2 x + \sinh^2 x$, with exponential growth for positive and negative values of x.

So the exponential growth of the real part of the function is governed by $\cosh x$, which blows up much slower than $\cosh^2 x + \sinh^2 x$. A plot of $\cosh x$ is shown over the same interval in Fig. 4.15 for comparison with Fig. 4.12. Be sure to compare the vertical axis of the two plots.

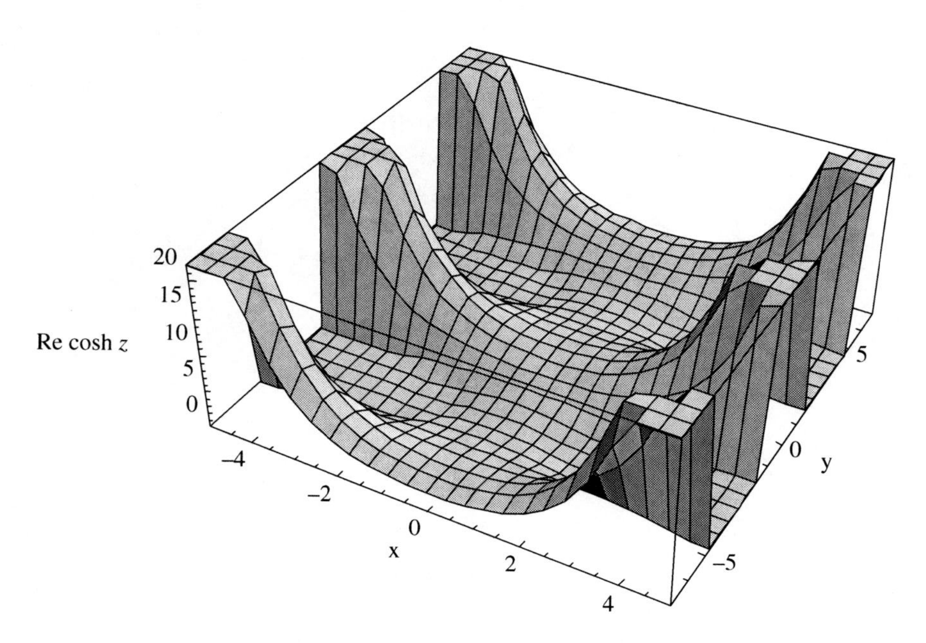

Figure 4.13 The real part of $\cosh z$.

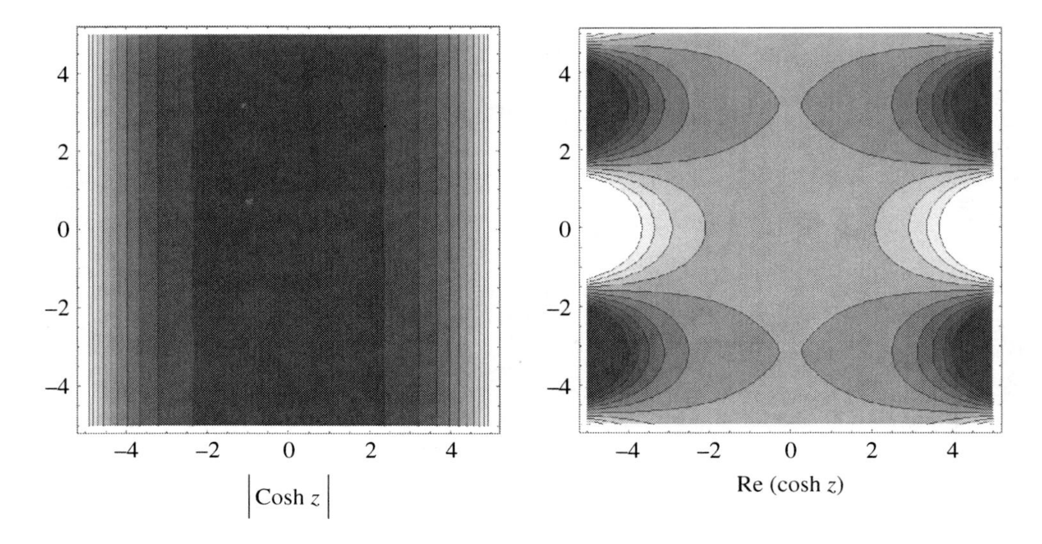

Figure 4.14 Contour plots of the modulus (on the left) and real part (on the right) of $\cosh z$.

Several relations exist which correlate the hyperbolic and trig functions for complex arguments. These include

$$\cosh iz = \cos z \tag{4.24}$$

$$\sinh iz = i \sin z \tag{4.25}$$

$$\cos iz = \cosh z \tag{4.26}$$

$$\sin iz = i \sinh z \tag{4.27}$$

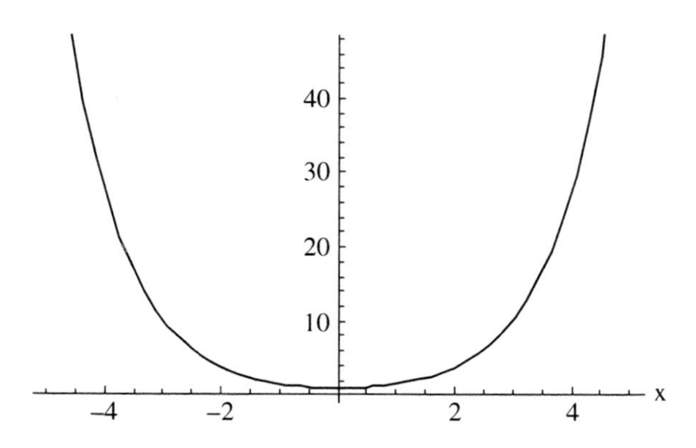

Figure 4.15 The real part of $\cosh z$ is influenced heavily by $\cosh x$.

These formulas are very easy to derive. For example:

$$\sin iz = \frac{e^{i(iz)} - e^{-i(iz)}}{2i} = -i\frac{e^{-z} - e^{z}}{2} = i\frac{e^{z} - e^{-z}}{2} = i\sinh z$$

The following identities, carried over from real variables, also hold

$$\cosh(-z) = \cosh z \tag{4.28}$$

$$\sinh(-z) = -\sinh z \tag{4.29}$$

$$\cosh^2 z - \sinh^2 z = 1 \tag{4.30}$$

$$\sinh(z+w) = \sinh z \cosh w + \cosh z \sinh w \tag{4.31}$$

$$\cosh(z+w) = \cosh z \cosh w + \sinh z \sinh w \tag{4.32}$$

The following identities incorporate trigonometric functions:

$$\sinh z = \sinh x \cos y + i \cosh x \sin y \tag{4.33}$$

$$\cosh z = \cosh x \cos y + i \sinh x \sin y \tag{4.34}$$

$$|\sinh z|^2 = \sinh^2 x + \sin^2 y \tag{4.35}$$

$$|\cosh z|^2 = \sinh^2 x + \cos^2 y \tag{4.36}$$

The hyperbolic functions are periodic. Looking at definitions in Eqs. (4.33) and (4.34), we see that this is due to the fact $\sinh z$ and $\cosh z$ that incorporate the trigonometric functions cosine and sin directly in their definitions. Therefore the period of the hyperbolic functions is given by

$$2\pi i \tag{4.37}$$

The zeros of the hyperbolic functions are given by

$$\cosh z = 0 \quad \text{if } z = \left(\frac{\pi}{2} + n\pi\right) \quad n = 0, \pm 1, \pm 2, \ldots \tag{4.38}$$

$$\sinh z = 0 \quad \text{if } z = n\pi i \quad n = 0, \pm 1, \pm 2, \ldots \tag{4.39}$$

We can also define the other hyperbolic functions analogous to the tangent, cosecant, and secant functions. In particular:

$$\tanh z = \frac{e^z - e^{-z}}{e^z + e^{-z}} \tag{4.40}$$

$$\operatorname{sech} z = \frac{2}{e^z + e^{-z}} \tag{4.41}$$

$$\operatorname{csch} z = \frac{2}{e^z - e^{-z}} \tag{4.42}$$

With the analogous identities

$$1 - \tanh^2 z = \operatorname{sech}^2 z \tag{4.43}$$

$$\tanh(z \pm w) = \frac{\tanh z \pm \tanh w}{1 \pm \tanh z \tanh w} \tag{4.44}$$

The hyperbolic functions also have inverses. Like the trigonometric functions, these inverses are defined using logarithms. Since the inverses are defined in terms of logarithms they are multivalued functions. These are given by

$$\cosh^{-1} z = \ln\left(z + \sqrt{z^2 - 1}\right) \tag{4.45}$$

$$\sinh^{-1} z = \ln\left(z + \sqrt{z^2 + 1}\right) \tag{4.46}$$

$$\tanh^{-1} z = \frac{1}{2} \ln\left(\frac{1+z}{1-z}\right) \tag{4.47}$$

Complex Exponents

Consider a function $f(z) = z^{\alpha}$, where α is a complex number. This function can be written in a convenient form using the exponential and natural log as follows:

$$f(z) = z^{\alpha} = e^{\alpha \ln z} \tag{4.48}$$

This can be generalized to the case when the exponent of a function is another complex function, that is,

$$f(z)^{g(z)} = e^{g(z)\ln[f(z)]} \tag{4.49}$$

From these definitions, we can see that powers of a complex variable z are multivalued functions.

EXAMPLE 4.7

Consider i^i and determine if it is multivalued.

SOLUTION

Using Eq. (4.48) we write

$$i^i = \exp(i \ln i)$$

Now

$$\ln i = \ln\left(1 \cdot e^{i\pi/2}\right) = \ln 1 + i\left(\frac{\pi}{2} + 2n\pi\right) = i\left(\frac{\pi}{2} + 2n\pi\right) \qquad \text{for } n = 0, \pm 1, \pm 2, \ldots$$

where Eq. (4.9) was used. So we have

$$i^i = \exp(i \ln i) = \exp\left(i\left[i\left(\frac{\pi}{2} + 2n\pi\right)\right]\right) = \exp\left(-\left(\frac{\pi}{2} + 2n\pi\right)\right)$$

where $n = 0, \pm 1, \pm 2, \ldots$, demonstrating that this is a multivalued function—in fact it has infinitely many values. Interestingly, they are all real numbers. Consider $n = 0$ for which $i^i = \exp(\pi/2) = 4.81$.

Derivatives of Some Elementary Functions

In Chap. 3, we have already studied derivatives in detail. In this section, we list some derivatives of the elementary functions for reference. Given a polynomial

$$f(z) = a_0 + a_1 z + a_2 z^2 + \cdots + a_n z^n$$

The derivative is given by

$$\frac{df}{dz} = a_1 + 2a_2 z + \cdots + n a_n z^{n-1}$$

The derivative of the exponential function e^z is

$$\frac{d}{dz} e^z = e^z \tag{4.50}$$

This result holds for the entire complex plane. Therefore the exponential function is analytic everywhere or we can say that it is entire.

The derivative of the logarithm is a bit more tricky. If we define

$$\ln z = \ln r + i\theta$$

where θ is restricted to the domain $\alpha < \theta < \alpha + 2\pi$, then we have a single-valued function with real and imaginary parts given by

$$u = \ln r \qquad v = \theta$$

These functions satisfy the Cauchy-Riemann equations, since

$$\frac{\partial u}{\partial r} = \frac{1}{r} \qquad \frac{\partial v}{\partial \theta} = 1$$

$$r\frac{\partial u}{\partial r} = \frac{\partial v}{\partial \theta}$$

and

$$\frac{\partial u}{\partial \theta} = -r\frac{\partial v}{\partial r} = 0$$

Given that the Cauchy-Riemann equations are satisfied, we can use Eq. (3.28), which stated that

$$f'(z) = e^{-i\theta}\left(\frac{\partial u}{\partial r} + i\frac{\partial v}{\partial r}\right)$$

So we've got

$$\frac{d}{dz}\ln z = e^{-i\theta}\left(\frac{\partial u}{\partial r} + i\frac{\partial v}{\partial r}\right) = e^{-i\theta}\left(\frac{1}{r} + i0\right) = \frac{1}{re^{i\theta}}$$

That is, the derivative of the natural logarithm for complex variables is the same as that in the calculus of real variables, namely:

$$\frac{d}{dz}\ln z = \frac{1}{z} \tag{4.51}$$

This result is valid when $|z| > 0$ and $\alpha < \arg z < \alpha + 2\pi$.

The derivatives of the trigonometric functions also correspond to the results we expect. Let's derive one example and then just state the other results. You can show that

$$\cos z = \cos(x + iy) = \cos x \cosh y - i \sin x \sinh y$$

So we have $u(x, y) = \cos x \cosh y$ and $v(x, y) = -\sin x \sinh y$. Then

$$\frac{\partial u}{\partial x} = -\sin x \cosh y \qquad \frac{\partial u}{\partial y} = \cos x \sinh y$$
$$\frac{\partial v}{\partial x} = -\cos x \sinh y \qquad \frac{\partial v}{\partial y} = -\sin x \cosh y$$

So it follows that

$$\frac{\partial u}{\partial x} = \frac{\partial v}{\partial y} \qquad \frac{\partial u}{\partial y} = -\frac{\partial v}{\partial x}$$

Since the Cauchy-Riemann equations are satisfied, we can write

$$\begin{aligned} f'(z) &= \frac{\partial u}{\partial x} + i\frac{\partial v}{\partial x} \\ &= -\sin x \cosh y - i \cos x \sinh y \end{aligned}$$

But $\sin z = \sin(x + iy) = \sin x \cosh y + i \cos x \sinh y$, therefore:

$$\frac{d}{dz}\cos z = -\sin z \tag{4.52}$$

You can also derive this very easily using the exponential representation of the sin and cosine functions. Other results can be derived similarly:

$$\frac{d}{dz}\sin z = \cos z \tag{4.53}$$

Since the exponential function is entire, the cosine and sin functions are also entire. Other derivatives follow from elementary calculus:

$$\frac{d}{dz}\tan z = \sec^2 z \qquad \frac{d}{dz}\cot z = -\csc^2 z \tag{4.54}$$

$$\frac{d}{dz}\sec z = \sec z \tan z \qquad \frac{d}{dz}\csc z = -\csc z \cot z \tag{4.55}$$

The derivatives of the hyperbolic functions can be derived easily using exponential representations:

$$\frac{d}{dz}\cosh z = \sinh z \qquad \frac{d}{dz}\sinh z = \cosh z \tag{4.56}$$

$$\frac{d}{dz}\tanh z = \operatorname{sech}^2 z \tag{4.57}$$

$$\frac{d}{dz}\operatorname{sech} z = -\operatorname{sech} z \tanh z \tag{4.58}$$

Finally, we note the derivative of a complex exponent:

$$\frac{d}{dz}z^{\alpha} = \alpha z^{\alpha-1} \tag{4.59}$$

Note, however, that since this is a multivalued function, this holds for $|z| > 0$, $0 < \arg z < 2\pi$ or some other interval.

Branches

A multivalued function repeats itself when z moves in a complete circle about the origin in the complex plane. When $0 \le \theta < 2\pi$, the function is single valued. We say that we are on one branch of the function. But as we let z traverse the circle again so we enter the region where $2\pi < \theta$, the function repeats. We say that we've entered another branch of the function. A multivalued function like this repeats itself any number of times.

For convenience, a barrier is set up at our choosing in the complex plane where we do not allow z to cross. This barrier is called a *branch cut*. The point from which the branch cut originates is called a *branch point*. The branch cut extends out from the branch point to infinity. For example, for a multivalued function, we can take the branch point to be the origin and the branch cut can extend out from the origin to positive infinity (Fig. 4.16).

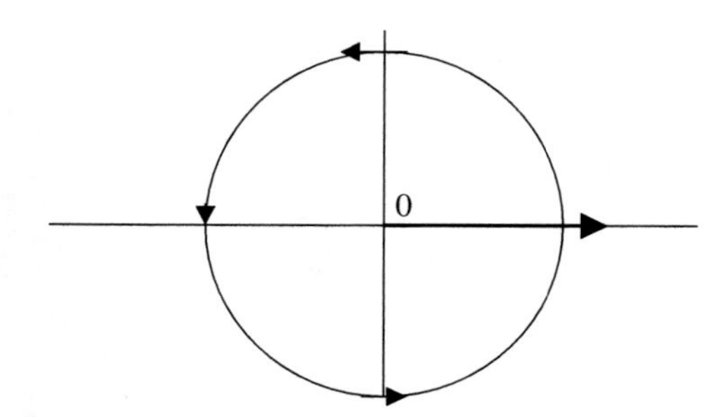

Figure 4.16 Some multivalued functions repeats themselves after z has completely gone around the origin. We prevent the function from being multivalued by staying on one branch. This means we cannot cross the branch cut, which we have chosen in this case to be the line from the origin to positive infinity. Note that a circle does not have to be used, we just have to let z go completely around the origin—a circle was used here for simplicity.

Summary

In this chapter, we described the basic properties of some elementary functions encountered in complex variables. These included polynomials, the complex exponential, the trig functions, the logarithm, the hyperbolic functions, and functions with complex exponents.

Quiz

1. Prove that $\cos(x+iy) = \cos x \cos iy - \sin x \sin iy$.
2. If $f(x) = e^x$, then f can never be negative. Is the same true of e^z?
3. Find a compact expression for $e^{2+3\pi i}$.
4. Find an identity for $1+\tan^2 z$ by using Eq. (4.12).
5. Find an identity for $\tan(z+w)$.
6. Are the inverse trig functions multivalued?

CHAPTER 5

Sequences and Series

It is common practice and often a necessity to represent a function of a real variable using an infinite series expansion. It turns out that this is also true when working with complex functions. As we will see, there are some new concepts involved when working with complex functions. We begin by considering *sequences.*

Sequences

Consider the positive integers $n = 1, 2, 3, \ldots$ and consider a function on the positive integers, which we denote by $f(n)$. We call such a function a sequence. The output of the function is a number: $f(n) = a_n$. So a sequence is an ordered set of numbers a_1, a_2, $a_3, \ldots$ and we refer to a_n as the nth term in the sequence. Sequences can also be indicated using curly braces, so we can write $\{f(n)\}$ or $\{a_n\}$.

THE LIMIT OF A SEQUENCE

It is desirable to find the limit of an infinite sequence to determine whether or not the sequence *converges* or approaches a specific finite value as n goes to infinity. Limits of sequences are defined in the standard way. Suppose that the limit of a sequence $f(n) = a_n$ is ℓ. Then this means that, given any positive number ε we can find a number N depending on ε such that

$$|a_n - \ell| < \varepsilon \qquad \text{for all } n > N \tag{5.1}$$

Using standard notation you are familiar with from calculus, we can write

$$\lim_{n\to\infty} a_n = \ell \tag{5.2}$$

If the limit of a sequence exists, we say that the sequence is *convergent*. If the limit does not exist or is infinite, then the sequence is *divergent*. Note that the limit of a given sequence is unique.

The limits of sequences satisfy all of the standard properties you are familiar with from your study of the limits of functions. Let us denote two sequences a_n and b_n such that $\lim_{n\to\infty} a_n = A$ and $\lim_{n\to\infty} b_n = B$. Then

$$\begin{aligned} \lim_{n\to\infty}(a_n \pm b_n) &= A \pm B \\ \lim_{n\to\infty} a_n b_n &= \left(\lim_{n\to\infty} a_n\right)\left(\lim_{n\to\infty} b_n\right) = AB \\ \lim_{n\to\infty} \frac{a_n}{b_n} &= \frac{\lim_{n\to\infty} a_n}{\lim_{n\to\infty} b_n} = \frac{A}{B} \end{aligned} \tag{5.3}$$

The last result holding provided that $B \neq 0$.

SEQUENCES OF COMPLEX FUNCTIONS

So far we haven't said anything about complex variables—we've just sketched out the notion of sequences in general. These ideas can be carried over to complex functions $f_n(z)$ defined on some region R of the complex plane. If $f(z)$ exists and is finite, and

$$\lim_{n\to\infty} f_n(z) = f(z) \tag{5.4}$$

on R we say that the sequence $f_n(z)$ converges to $f(z)$ on R. The formal definition of this limit follows from Eq. (5.1). That is, given any $\varepsilon > 0$ there exists an N depending on ε such that

$$|f_n(z) - f(z)| < \varepsilon \tag{5.5}$$

for $n > N$. If the limit does not exist or is infinite, then the sequence is divergent.

EXAMPLE 5.1

Let a sequence z_n be defined as

$$z_n = \frac{2}{n^2} + 3i$$

Does this sequence converge? Find an N such that Eq. (5.5) is satisfied.

SOLUTION

We examine the limit of this sequence. We have

$$\lim_{n\to\infty} z_n = \lim_{n\to\infty} \frac{2}{n^2} + 3i = \lim_{n\to\infty} \frac{2}{n^2} + \lim_{n\to\infty} 3i = 0 + 3i = 3i$$

That is, the sequence converges and its limit is $3i$. Formally, given any $\varepsilon > 0$ we need to find an N such that $|(2/n^2) + 3i - 3i| = |2/n^2| < \varepsilon$ for $n > N$. So we have

$$\left|\frac{2}{n^2}\right| < \varepsilon \Rightarrow n > \sqrt{\frac{2}{\varepsilon}}$$

This means that Eq. (5.5) is satisfied if we take

$$N = \sqrt{\frac{2}{\varepsilon}}$$

Note that since N depends on ε, the sequence is not uniformly convergent (see Sec. "Uniformly Converging Series" later in the chapter).

Infinite Series

By summing up the individual terms in a sequence we can construct a series. This can be done using so-called *partial sums*. That is, let $\{a_n(z)\}$ be some complex sequence. Then we can form partial sums as follows:

$$
\begin{aligned}
S_1 &= a_1(z) \\
S_2 &= a_1(z) + a_2(z) \\
&\vdots \\
S_n &= a_1(z) + a_2(z) + \cdots + a_n(z)
\end{aligned}
$$

So, the nth partial sum is constructed by adding up the first n terms of the sequence. If we let $n \to \infty$, we obtain an infinite series:

$$\sum_{n=1}^{\infty} a_n(z) \tag{5.6}$$

If the following condition holds:

$$\lim_{n\to\infty} S_n(z) = S(z) \tag{5.7}$$

where $S(z)$ is a finite quantity we say that the series is *convergent*. If Eq. (5.7) does not hold then the series is divergent. A necessary but not sufficient condition for a series to be convergent is that the following condition holds:

$$\lim_{n\to\infty} a_n(z) = 0 \tag{5.8}$$

In the next section, we'll review some tests that can be used to determine whether or not a series converges.

Convergence

An important concept used in working with series in complex analysis is the *radius of convergence R*. Simply put, we want to know over what region R of the complex plane does the series converge. It may be that the series converges everywhere, or it could turn out that the series only converges inside the unit disc, say.

First let's take a look at sequences again. If each term in a sequence is larger than or equal to the previous term, which means that

$$a_{n+1} \geq a_n$$

We say that the sequence is *monotonic increasing*. On the other hand, if

$$a_{n+1} \leq a_n$$

then the sequence is *monotonic decreasing*. If each term in the sequence is bounded above by some constant M:

$$|a_n| < M \tag{5.9}$$

then we say that the sequence is *bounded*. A bounded monotonic sequence (either increasing or decreasing) converges.

CAUCHY'S CONVERGENCE CRITERION

Saying that a sequence converges is the same as saying that it has a limit, so we can formalize the notion of convergence. Leave it up to Cauchy to have done that for us. So, $\{a_n\}$ converges if given an $\varepsilon > 0$ we can find an N such that

$$|a_m - a_n| < \varepsilon \qquad \text{for } m, n > N$$

Cauchy's convergence criterion is necessary and sufficient to show convergence of a sequence.

CONVERGENCE OF A COMPLEX SERIES

Remember that any complex function $f(z)$ can be written in terms of real and imaginary parts, just like a complex number. The real and imaginary parts are themselves real functions. So one way to check convergence is to check the convergence of the real and imaginary parts—assuming we have a series representation available—and seeing if they converge. So a necessary and sufficient condition that a series of the form $\Sigma a_j + ib_j$ converges is that the two series Σa_j and Σb_j both converge.

Convergence Tests

The following convergence tests can be used to evaluate whether or not a series converges. If we say that a series $\sum a_n$ converges *absolutely,* we mean that

$$\sum_{n=1}^{\infty} |a_n| \tag{5.10}$$

converges. The first test that we can apply for convergence is the *comparison test.* The comparison test tells us that if $\sum |b_n|$ converges and $|a_n| \le |b_n|$, then the series $\sum |a_n|$ converges absolutely. If $\sum |b_n|$ diverges and $|a_n| \ge |b_n|$, the series $\sum |a_n|$ also diverges. However, we can't say anything about the series $\sum a_n$.

The *ratio test* is a nice test that appeals to common sense. We take the ratio of the terms a_{n+1} to a_n and take the limit $n \to \infty$. Let

$$\lim_{n\to\infty} \left| \frac{a_{n+1}}{a_n} \right| = R \tag{5.11}$$

There are two possibilities:

- If $R < 1$ then the series converges absolutely.
- If $R > 1$ then the series is divergent.

If $R = 1$ then no information is available from the test.

The *n*th *root test* checks the limit:

$$\lim_{n\to\infty} \sqrt[n]{|a_n|} = R \tag{5.12}$$

The possibilities here are the same we encountered with the ratio test. These are

- If $R < 1$ then the series converges absolutely.
- If $R > 1$ then the series is divergent.

If $R = 1$ then no information is available from the test.

Raabe's test checks the limit:

$$\lim_{n\to\infty} n\left(1-\left|\frac{a_{n+1}}{a_n}\right|\right) \tag{5.13}$$

Again

- If $R<1$ then the series converges absolutely.
- If $R>1$ then the series is divergent.

If $R=1$ then no information is available from the test.

Finally, we consider the *Weierstrass M-test.* Suppose that $|a_n(z)| \le M_n$. If M_n does not depend on z in some region of the complex plane where $|a_n(z)| \le M_n$ holds and $\Sigma\, M_n$ converges, then $\Sigma\, a_n(z)$ is uniformly convergent.

Uniformly Converging Series

We often find in the limits we compute that N depends on ε. When a series is uniformly convergent, then for any $\varepsilon > 0$ there is an N not depending on ε such that $|a_n(z) - R| < \varepsilon$ for $n > N$, where R is the limit. That is, if the same N holds for all z in a given region D of the complex plane, then we say that the convergence is *uniform*.

Power Series

A series that can be written as

$$a_0 + a_1 z + a_2 z^2 + \cdots = \sum_{n=0}^{\infty} a_n (z - z_0)^n \tag{5.14}$$

where the a_n are constants is called a *power series.* When the series converges for $|z - z_0| < R$ we say that R is the *radius of convergence.* The series diverges if $|z - z_0| > R$. For $|z - z_0| = R$, the series may converge or it may diverge. Often in complex analysis, the region over which the series converges is a disc so the term radius has a literal geometric interpretation.

Taylor and Maclaurin Series

Suppose that a complex function $f(z)$ is analytic in some region of the complex plane and let z_0 be a point inside that region. Then $f(z)$ has a power series expansion with expansion coefficients calculated by computing derivatives of the function at that point, giving the *Taylor series expansion* of the function:

$$\begin{aligned} f(z) &= f(z_0) + f'(z_0)(z - z_0) + \frac{f''(z_0)}{2!}(z - z_0)^2 \\ &\quad + \cdots + \frac{f^{(n)}(z_0)}{n!}(z - z_0)^n + \cdots \end{aligned} \tag{5.15}$$

If we set $z_0 = 0$, that is take the series expansion about the origin, we have a *Maclaurin series.*

Theorems on Power Series

The most important fact about a convergent power series you should file away in your mind is that within the radius of convergence, you can differentiate a power series term by term, or you can integrate it term by term along any curve that lies within its radius of convergence (see Chaps. 6 and 7).

EXAMPLE 5.2

Find the Taylor expansion of $f(z) = 1/(1 - z + z^2)$ about the origin.

SOLUTION

We will calculate the first two derivatives. First, note that

$$f(0) = \frac{1}{1 - 0 + 0^2} = 1$$

The first derivative is

$$f'(z) = \frac{d}{dz}\frac{1}{1 - z + z^2} = -\frac{1}{(1 - z + z^2)^2}(-1 + 2z)$$

$$\Rightarrow f'(0) = 1$$

The second derivative is

$$f''(z) = \frac{d}{dz}\left(-\frac{1}{(1-z+z^2)^2}(-1+2z)\right)$$
$$= \frac{2}{(1-z+z^2)^3}(-1+2z)^2 - \frac{2}{(1-z+z^2)^2}$$
$$f''(0) = 0$$

Finally, the third derivative is

$$f'''(z) = \frac{d}{dz}\left[\frac{2}{(1-z+z^2)^3}(-1+2z)^2 - \frac{2}{(1-z+z^2)^2}\right]$$
$$= -\frac{6(-1+2z)^3}{(1-z+z^2)^4} + \frac{12(-1+2z)}{(1-z+z^2)^3}$$
$$f''(0) = -6$$

So we have

$$f(z) = f(0) + z\,f'(0) + \frac{z^2}{2!}f''(0) + \frac{z^3}{3!}f'''(0) + \cdots$$
$$= 1 + z - z^3 + \cdots$$

EXAMPLE 5.3

Use the Weierstrass M test to determine whether or not the series $\Sigma\, a_n \cos nx + b_n \sin nx$ converges, provided that the series $\Sigma|a_n|$ and $\Sigma|b_n|$ converge and if $x \in [-\pi, \pi]$.

SOLUTION

The values of $\cos nx$ and $\sin nx$ may be positive, negative, or zero. However, we know that they are bounded by 1, that is for all n:

$$|\cos nx| \le 1 \qquad \text{and} \qquad |\sin nx| \le 1$$

It follows that

$$a_n|\cos nx| \le a_n \qquad \text{and} \qquad b_n|\sin nx| \le b_n$$

Now, since the series $\sum |a_n|$ and $\sum |b_n|$ converge, given $\varepsilon > 0$ and $n \geq m$:

$$\left| |a_m| + |b_m| + |a_{m+1}| + \cdots + |a_n| + |b_n| \right| < \varepsilon$$

We then have

$$\begin{aligned} &|a_m \cos mx + b_m \sin mx + \cdots + a_n \cos nx + b_n \sin nx| \\ &\leq \left| |a_m| + |b_m| + |a_{m+1}| + \cdots + |a_n| + |b_n| \right| < \varepsilon \end{aligned}$$

We have thus constructed a series of numbers that converges $\sum a_n + b_n$ for which $a_k + b_k \geq |a_k \cos kx + b_k \sin kx|$ for all $x \in [-\pi, \pi]$. By the Weierstrass M test, $\sum a_n \cos nx + b_n \sin nx$ is uniformly convergent. Since $\cos nx$ and $\sin nx$ are periodic with period 2π, the series is uniformly convergent for $-\infty < x < \infty$.

Some Common Series

There are many functions which are encountered over and over again in analysis and applied mathematics. You should be familiar with their power series representations. Some of the functions we take note of and their Taylor expansions are

$$e^z = 1 + z + \frac{1}{2!}z^2 + \cdots + \frac{1}{n!}z^n + \cdots = \sum_{n=0}^{\infty} \frac{1}{n!} z^n \tag{5.16}$$

$$\sin z = z - \frac{z^3}{3!} + \frac{z^5}{5!} - \frac{z^7}{7!} + \cdots + \frac{(-1)^{n-1}}{(2n-1)!} z^{2n-1} + \cdots = \sum_{n=1}^{\infty} \frac{(-1)^{n-1}}{(2n-1)!} z^{2n-1} \tag{5.17}$$

$$\cos z = 1 - \frac{z^2}{2!} + \frac{z^4}{4!} - \frac{z^6}{6!} + \cdots + \frac{(-1)^n}{(2n)!} z^{2n} + \cdots = \sum_{n=1}^{\infty} \frac{(-1)^n}{(2n)!} z^{2n} \tag{5.18}$$

$$\ln(1+z) = z - \frac{z^2}{2} + \frac{z^3}{3} + \cdots + \frac{(-1)^{n-1}}{n} z^n + \cdots = \sum_{n=1}^{\infty} \frac{(-1)^{n-1}}{n} z^n \tag{5.19}$$

$$\tan^{-1} z = z - \frac{z^3}{3} + \frac{z^5}{5} - \cdots + \frac{(-1)^{n-1}}{2n-1} z^{2n-1} + \cdots = \sum_{n=1}^{\infty} \frac{(-1)^{n-1}}{2n-1} z^{2n-1} \tag{5.20}$$

If $|r| < 1$, then the *geometric series* converges as

$$\sum_{n=0}^{\infty} r^n = \frac{1}{1-r} \tag{5.21}$$

The *harmonic series* is divergent:

$$\sum_{n=1}^{\infty} \frac{1}{n} = \infty \tag{5.22}$$

But the *alternating harmonic series* is convergent:

$$\sum_{n=1}^{\infty} \frac{(-1)^{n-1}}{n} = \ln 2 \tag{5.23}$$

EXAMPLE 5.4

A *Bessel function* is one that solves the differential equation $x^2(d^2y/dx^2) + x(dy/dx) + (x^2 - a^2)y = 0$.

The series representation of the Bessel function is given by $J_0(x) = \Sigma_{n=0}^{\infty}[\{(-1)^n/(n!)^{2^{\cdot}}\}(x/2)^2]$.
Show that we can write:

$$J_0(x) = \frac{1}{2\pi}\int_0^{2\pi} \cos(x\cos(\phi))\, d\phi$$

SOLUTION

We use the series representation of the cosine function:

$$\begin{aligned}
\frac{1}{2\pi}\int_0^{2\pi} \cos(x\cos(\phi))\, d\phi &= \frac{1}{2\pi}\int_0^{2\pi} \sum_{n=0}^{\infty} \frac{(-1)^n}{(2n)!}(x\cos\phi)^{2n}\, d\phi \\
&= \sum_{n=0}^{\infty} \frac{1}{2\pi}\int_0^{2\pi} \frac{(-1)^n}{(2n)!}(x\cos\phi)^{2n}\, d\phi \\
&= \sum_{n=0}^{\infty} \frac{1}{2\pi}\frac{(-1)^n}{(2n)!}x^{2n}\int_0^{2\pi} (\cos\phi)^{2n}\, d\phi \\
&= \frac{1}{2\pi}\sum_{n=0}^{\infty} \frac{(-1)^n}{(2n)!}x^{2n}\int_0^{2\pi} (\cos\phi)^{2n}\, d\phi
\end{aligned}$$

You can verify that

$$\int_0^{2\pi} (\cos\phi)^{2n} d\phi = \frac{(2n)!}{2^{2n}(n!)^2} 2\pi$$

Hence

$$\begin{aligned}
\frac{1}{2\pi}\int_0^{2\pi} \cos(x\cos(\phi))\,d\phi &= \frac{1}{2\pi}\sum_{n=0}^{\infty}\frac{(-1)^n}{(2n)!}x^{2n}\int_0^{2\pi}(\cos\phi)^{2n}\,d\phi \\
&= \frac{1}{2\pi}\sum_{n=0}^{\infty}\frac{(-1)^n}{(2n)!}x^{2n}\frac{(2n)!}{2^{2n}(n!)^2}2\pi \\
&= \sum_{n=0}^{\infty}\frac{(-1)^n}{(n!)^2}\frac{x^{2n}}{2^{2n}} = \sum_{n=0}^{\infty}\frac{(-1)^n}{(n!)^2}\left(\frac{x}{2}\right)^{2n} = J_0(x)
\end{aligned}$$

EXAMPLE 5.5

Given that

$$\sinh x = \frac{e^x - e^{-x}}{2}$$

find a series representation for $\sinh^{-1} x$.

SOLUTION

The Maclaurin theorem can be used to write a series representation of $\sinh x$. This is given by

$$\sinh x = x + \frac{x^3}{3!} + \frac{x^5}{5!} + \frac{x^7}{7!} + \cdots = \sum_{n=0}^{\infty}\frac{1}{(2n+1)!}x^{2n+1}$$

The inverse will have some series expansion which we write as

$$\sinh^{-1} x = b_0 + b_1 x + b_2 x^2 + b_3 x^3 + \cdots$$

We label the coefficients in the series expansion of sinh by a_j. We find that

$$b_0 = a_0 = 0$$

$$b_1 = \frac{1}{a_1} = 1 \qquad b_2 = -\frac{a_2}{a_1^3} = 0$$

$$b_3 = \frac{1}{a_1^5}\left(2a_2^2 - a_1 a_3\right) = -\frac{1}{6}$$

Therefore it follows that

$$\sinh^{-1} x = x - \frac{1}{6}x^3 + \cdots$$

EXAMPLE 5.6

Find a series expansion of $f(z) = (1+z)^k$ about $z = 0$.

SOLUTION

We seek a series representation of the form:

$$f(z) = \sum_{n=0}^{\infty} (z - z_0)^n \frac{f^n(z_0)}{n!}$$

Taking $z_0 = 0$, we have the following relations:

$$\begin{aligned}
&f(0) = 1, \\
&f'(z) = k(1+z)^{k-1} && \Rightarrow f'(0) = k \\
&f''(z) = k(k-1)(1+z)^{k-2} && \Rightarrow f''(0) = k(k-1) \\
&f'''(z) = k(k-1)(k-2)(1+z)^{k-3} && \Rightarrow f'''(0) = k(k-1)(k-2)
\end{aligned}$$

At $z_0 = 0$ the series representation is

$$\begin{aligned}
f(z) &= \sum_{n=0}^{\infty} (z - z_0)^n \frac{f^n(z_0)}{n!} \\
&= f(0) + \left.\frac{df}{dz}\right|_{z=0} + \left.\frac{1}{2!}\frac{d^2 f}{dz^2}\right|_{z=0} + \left.\frac{1}{3!}\frac{d^3 f}{dz^3}\right|_{z=0} + \cdots \\
&= 1 + kx + \frac{1}{2}k(k-1)z^2 + \frac{k(k-1)(k-2)}{6}z^3 + \cdots \\
&= \sum_{n=0}^{\infty} \binom{k}{n} z^n
\end{aligned}$$

EXAMPLE 5.7

Find a series representation of $f(z) = \cos z$ about the point $z = \pi/4$.

SOLUTION

While we could do a Taylor expansion, a little algebraic manipulation will give the same result. We can find a series representation of this function by first recalling that

$$\cos(a+b) = \cos a \cos b - \sin a \sin b$$

Now let $z = w + \pi/4$. Then we have

$$\begin{aligned} f(z) = \cos z &= \cos(w+\pi/4) \\ &= \cos w \cos \pi/4 - \sin w \sin \pi/4 \\ &= \frac{1}{\sqrt{2}}(\cos w - \sin w) \end{aligned}$$

Expanding each trigonometric function we get

$$\begin{aligned} f(z) &= \frac{1}{\sqrt{2}}\left(1 - \frac{w^2}{2!} + \frac{w^4}{4!} + \cdots - \left(w - \frac{w^3}{3!} + \frac{w^5}{5!} - \cdots\right)\right) \\ &= \frac{1}{\sqrt{2}}\left(1 - w - \frac{w^2}{2!} + \frac{w^3}{3!} + \frac{w^4}{4!} - \frac{w^5}{5!} + \cdots\right) \\ &= \frac{1}{\sqrt{2}}\left(1 - (z-\pi/4) - \frac{(z-\pi/4)^2}{2!} + \frac{(z-\pi/4)^3}{3!} + \frac{(z-\pi/4)^4}{4!} - \frac{(z-\pi/4)^5}{5!} + \cdots\right) \end{aligned}$$

EXAMPLE 5.8

Find the disc of convergence for $\Sigma_{n=1}^{\infty}[(n!z^n)/n^n]$.

SOLUTION

We can find the disc of convergence for this series by using the ratio test. We have

$$a_{n+1} = \frac{(n+1)!z^{n+1}}{(n+1)^{n+1}} \qquad a_n = \frac{n!z^n}{n^n}$$

Therefore the ratio of the $(n + 1)$ term to the nth term is

$$\frac{a_{n+1}}{a_n} = \frac{(n+1)!z^{n+1}}{(n+1)^{n+1}} \Big/ \frac{n!z^n}{n^n} = \frac{(n+1)!z^{n+1}}{(n+1)^{n+1}} \frac{n^n}{n!z^n}$$

Now recall that

$$(n+1)! = (n+1)n(n-1)(n-2)\cdots 2\cdot 1 = (n+1)n!$$

So the ratio simplifies to

$$\begin{aligned}\frac{a_{n+1}}{a_n} &= \frac{(n+1)!\,z^{n+1}}{(n+1)^{n+1}}\frac{n^n}{n!\,z^n} = \frac{(n+1)z^{n+1}}{(n+1)^{n+1}}\frac{n^n}{z^n} \\ &= \frac{(n+1)z\,n^n}{(n+1)^{n+1}} \\ &= \frac{z\,n^n}{(n+1)^n} = \frac{zn^n}{n^n\left(1+\dfrac{1}{n}\right)^n} = \frac{z}{\left(1+\dfrac{1}{n}\right)^n}\end{aligned}$$

Recalling that

$$\lim_{n\to\infty}\left(1+\frac{1}{n}\right)^n = e \tag{5.24}$$

The ratio test in this case becomes

$$\lim_{n\to\infty}\left|\frac{z}{\left(1+\dfrac{1}{n}\right)^n}\right| = \lim_{n\to\infty}\frac{|z|}{\left(1+\dfrac{1}{n}\right)^n} = \frac{|z|}{e}$$

Therefore the series converges when

$$\frac{|z|}{e} < 1$$

and diverges when

$$\frac{|z|}{e} > 1$$

In other words, the series is convergent if $|z| < e$, so the radius of convergence is given by $R = e$.

EXAMPLE 5.9

Consider the series $\sum_{n=1}^{\infty}[z^n/\{n(\log n)^a\}]$, where $a > 0$. Determine the radius of convergence.

SOLUTION

To find the radius of convergence for this series we use the root test:

$$\lim_{n\to\infty}\sqrt[n]{|a_n|} = L$$

In this case we've got

$$\begin{aligned}\lim_{n\to\infty}\sqrt[n]{|a_n|} &= \lim_{n\to\infty}\sqrt[n]{n(\log n)^a}\\ &= \lim_{n\to\infty}\exp\left[\sqrt[n]{n(\log n)^a}\right]\\ &= \lim_{n\to\infty}\exp\left[\frac{1}{n}\log n + \frac{a}{n}\log(\log n)\right] = e^0 = 1\end{aligned}$$

Hence the radius of convergence is $R = 1$.

EXAMPLE 5.10

Describe the convergence of the series:

$$\sum_{n=1}^{\infty}\frac{z^n}{n^2(1-z^n)}$$

SOLUTION

First, consider the case where $z^n = 1$. It is clear that this will cause the series to blow up. This means that the nth roots of unity are not permitted for this series, that is

$$z \neq e^{2\pi ik/n} \qquad \text{for } n \geq 1 \qquad k = 0,1,2,\ldots,n-1$$

So we conclude the series is divergent for $|z| = 1$. Now we check the case of $|z| < 1$. Notice that since $|z| < 1$:

$$\left|\frac{z^n}{1-z^n}\right| = \left|\frac{z^n}{z^n\left(\frac{1}{z^n}-1\right)}\right| = \left|\frac{1}{\frac{1}{z^n}-1}\right| < 1$$

The series

$$S = \sum_{n=1}^{\infty} \frac{1}{n^2}$$

is convergent. We have

$$\left| \frac{z^n}{n^2(1-z^n)} \right| \leq \frac{1}{n^2}$$

Since $|z^n/(1-z^n)| < 1$. Therefore by the Weierstrass M test, $\Sigma_{n=1}^{\infty} [z^n/\{n^2(1-z^n)\}]$ is convergent absolutely inside the unit disc.

Finally, we consider the case where $|z| > 1$. It is easy to see that the series converges in this case since

$$\left| \frac{z^n}{1-z^n} \right| = \left| \frac{1}{\frac{1}{z^n} - 1} \right| \to 1 \text{ as } |z|^n \to \infty$$

EXAMPLE 5.11

Describe the convergence of the series $F(z) = \Sigma_{n=1}^{\infty} [(-1)^{n-1}/(n+|z|)]$.

SOLUTION

Notice that since the series contains $|z|$ and not z, the series is actually a series of real numbers. Suppose that we pick some arbitrary $z \in \mathbb{C}$. Then we can pick a k that satisfies

$$k \leq |z| < k+1$$
$$\Rightarrow n+k \leq n+|z| < n+k+1$$

Which means that

$$\frac{1}{n+k+1} < \frac{1}{n+|z|} \leq \frac{1}{n+k}$$

It follows that

$$\sum_{n=1}^{\infty} \frac{1}{n+k+1} = \sum_{m=k+2}^{\infty} \frac{1}{m} < \sum_{n=1}^{\infty} \frac{1}{n+|z|}$$

Now, $\sum_{m=k+2}^{\infty}(1/m)$ is the tail of the harmonic series, which is divergent. Therefore the series $\sum_{n=1}^{\infty}[(-1)^{n-1}/(n+|z|)]$ does not converge absolutely. However, the series does converge. Notice that

$$\lim_{n\to\infty}\frac{(-1)^{n-1}}{n+|z|}=0$$

Furthermore, it is the case that

$$|a_{n+1}|=\frac{1}{n+1+|z|}\leq|a_n|=\frac{1}{n+|z|}$$

Since $\lim_{n\to\infty}a_n=0$, the series converges (but not absolutely, as we've already established). Now we investigate whether or not it converges uniformly. Consider the sum

$$F(z)-F_{2N-1}(z)=\sum_{n=2N}^{\infty}\frac{(-1)^{n-1}}{n+|z|}$$

Let $n=2k,\Rightarrow k=n/2$ and for $n=2N,k=N$. So we can write

$$\begin{aligned}F(z)-F_{2N-1}(z)&=\sum_{k=N}^{\infty}\frac{(-1)^{2k-1}}{2k+|z|}\\&=\sum_{k=N}^{\infty}\left(\frac{1}{2k+1+|z|}-\frac{1}{2k+|z|}\right)\\&=-\sum_{k=N}^{\infty}\frac{1}{(2k+1+|z|)(2k+|z|)}\\&\leq\sum_{k=N}^{\infty}\frac{1}{(2k+1)(2k)}<\sum_{n=2N}^{\infty}\frac{1}{n(n+1)}=\frac{1}{N}\end{aligned}$$

Therefore, it is possible to choose a positive integer M such that

$$|F(z)-F_n(z)|<\varepsilon \qquad \text{for all } n>M$$

So the series converges uniformly.

EXAMPLE 5.12

Let the domain of definition D be the unit disc and show that $\sum_{n=1}^{\infty} nz^n = z/(1-z)^2$.

SOLUTION

You can check to see if the series converges inside the unit disc. Since it does, we can differentiate it term by term. Let's recall the geometric series in Eq. (5.21):

$$\sum_{n=0}^{\infty} r^n = \frac{1}{1-r}$$

Notice what happens if we take the derivative with respect to r of both sides:

$$\frac{d}{dr}\sum_{n=0}^{\infty} r^n = \sum_{n=0}^{\infty}\frac{d}{dr} r^n = \sum_{n=1}^{\infty} nr^{n-1}$$

$$\frac{d}{dr}\frac{1}{1-r} = \frac{1}{(1-r)^2}$$

This demonstrates that

$$\sum_{n=1}^{\infty} nr^{n-1} = \frac{1}{(1-r)^2}$$

The geometric series is convergent provided that $|r| < 1$. In the complex plane, this is the same as saying that z lies in the unit disc. Hence

$$\sum_{n=1}^{\infty} nz^{n-1} = \frac{1}{(1-z)^2}$$

Now multiply both sides by z to obtain the desired result:

$$\sum_{n=1}^{\infty} nz^n = \frac{z}{(1-z)^2}$$

Laurent Series

A *Laurent series* is a serial representation of a function of a complex variable $f(z)$. A major difference you will notice when comparing a Laurent series to a Taylor series or power series expansion is that a Laurent series includes terms with negative powers.

In principle, the powers can range all the way down to $-\infty$, but in many if not most cases only a few terms with negative power are included. So, generally speaking, the Laurent series of a complex function $f(z)$ about the point $z = z_0$ is given by

$$f(z) = \sum_{n=-\infty}^{\infty} a_n (z - z_0)^n \tag{5.25}$$

The coefficients in the expansion are calculated using Cauchy's integral formula, which we discuss in the Chaps. 6 and 7. Stating it for the record:

$$a_n = \frac{1}{2\pi i} \oint \frac{f(z)dz}{(z - z_0)^{n+1}} \qquad \text{for } n = 0,1,2,\ldots \tag{5.26}$$

The integral is taken along curves defining an annulus enclosing the point z_0. In Eq. (5.26), the curve used for the integration is the outer curve defining the annulus. The negative coefficients in the series are calculated using

$$a_n = \frac{1}{2\pi i} \oint f(z)\,(z - z_0)^{n-1}\,dz \qquad \text{for } n = 1,2,3,\ldots \tag{5.27}$$

In this case the inner curve is used (see Fig. 5.1). By the deformation of path theorem, we know that we can use any concentric circle enclosing the singular point z_0 to calculate the integral. As a result, formula in Eq. (5.26) is universally valid for $n = 0, \pm 1, \pm 2, \ldots$

A Laurent series can be written in the form

$$f(z) = a_0 + a_1(z - z_0) + a_2(z - z_0)^2 + \cdots + \frac{a_{-1}}{z - z_0} + \frac{a_{-2}}{(z - z_0)^2} + \cdots \tag{5.28}$$

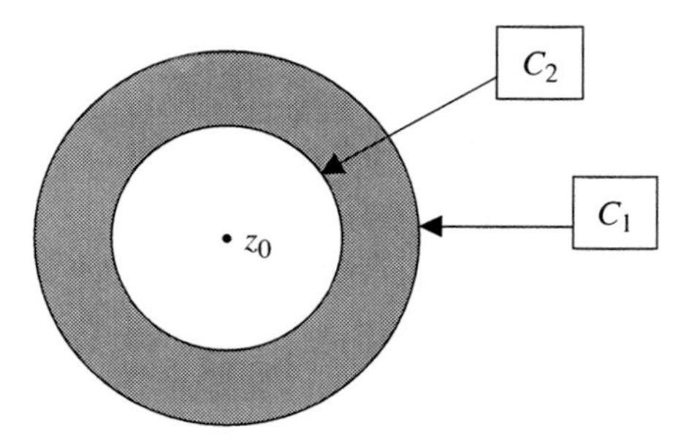

Figure 5.1 An illustration of an annular region used for integration in the determination of the coefficients of a Laurent expansion.

The summation including the terms with negative indices is called the *principal part* of the series:

$$\frac{a_{-1}}{z-z_0}+\frac{a_{-2}}{(z-z_0)^2}+\cdots \tag{5.29}$$

We call the points z_0 that give rise to terms with inverse powers of $z-z_0$ in a Laurent expansion *singular points* or *singularities.* Colloquially speaking, singularities represent points at which a function will blow up. It is also a point at which the function is not differentiable.

The *analytic part* of the series is given by the part of the expansion, which resembles an ordinary power series expansion:

$$a_0+a_1(z-z_0)+a_2(z-z_0)^2+\cdots \tag{5.30}$$

Types of Singularities

The first type of singularity we encounter is called a *removable singularity,* because it is a point $z=z_0$ at which the function *appears* to blow up, but at which a formal calculation $\lim_{z\to z_0} f(z)$ exists. The quintessential example (which we will remind you of again in Chap. 7) is $f(z)=(\sin z)/z$. The value $f(0)$ is not defined, but $\lim_{z\to} f(z)=1$. If you grasp this then you understand the concept of the removable singularity.

Suppose that the principal part of a Laurent series only has a finite number of terms:

$$\frac{a_{-1}}{z-z_0}+\frac{a_{-2}}{(z-z_0)^2}+\cdots+\frac{a_{-n}}{(z-z_0)^n} \tag{5.31}$$

Then the point $z=z_0$ is called a *pole of order n.* A pole causes the function to blow up at $z=z_0$. If a_{-1} is the only nonzero coefficient in the principal part of the series, we say that $z=z_0$ is a *simple pole.*

An *essential singularity* is one which results in an infinite number of inverse power terms in the Laurent expansion. That is, the principal part of the Laurent expansion is nonterminating.

A *branch point* $z=z_0$ is a point of a multivalued function where the function changes value when a curve winds once around z_0.

A *singularity at infinity* is a zero of $f(z)$ if we let $z = 1/w$ and then consider the function $F(w) = f(1/z)$.

Entire Functions

We first met the concept of an entire function in Chap. 3. Now that we have introduced the concept of a Laurent series, we have a systematic way to determine if a function is entire. An entire function is analytic throughout the entire complex plane. The Laurent expansion of an entire function cannot contain a principal part. Or expressed another way, an entire function has a Taylor series expansion with an infinite radius of convergence. The radius of convergence is infinite since the function is analytic on the entire complex plane.

Meromorphic Functions

A *meromorphic* function is analytic everywhere in the complex plane *except at a finite number of poles.*

EXAMPLE 5.13

Describe the singularities of $f(z) = 1/[(z-2)(z+4)^3]$. Is this function entire?

SOLUTION

The function $f(z)$ has singularities at $z = 2$ and $z = -4$. The pole at $z = 2$ is a simple pole because the power of this term is -1. The pole at $z = -4$ is a pole of order 3.

The function is not entire, because it is not analytic at the poles. Since there are a finite number of poles, the function is meromorphic.

EXAMPLE 5.14

Suppose that $f(z) = (z-1)\cos[1/(z+2)]$. Find the Laurent expansion of this function about the point $z = -2$ and describe the nature of any singularities. Identify the analytic and principal parts of the series expansion.

SOLUTION

Recall the power series expansion of the cosine function:

$$\cos z = 1 - \frac{1}{2}z^2 + \frac{1}{4!}z^4 - \frac{1}{6!}z^6 + \cdots$$

Now let $w = z + 2$ to simplify notation. Then

$$\cos\left(\frac{1}{z+2}\right) = \cos\left(\frac{1}{w}\right) = 1 - \frac{1}{2}\left(\frac{1}{w}\right)^2 + \frac{1}{4!}\left(\frac{1}{w}\right)^4 - \frac{1}{6!}\left(\frac{1}{w}\right)^6 + \cdots$$

The term $z - 1 = w - 3$ and so

$$\begin{aligned} f(z) &= (z-1)\cos\left(\frac{1}{z+2}\right) = (w-3)\cos\left(\frac{1}{w}\right) \\ &= (w-3)\left(1 - \frac{1}{2}\left(\frac{1}{w}\right)^2 + \frac{1}{4!}\left(\frac{1}{w}\right)^4 - \frac{1}{6!}\left(\frac{1}{w}\right)^6 + \cdots\right) \\ &= w - 3 - w\frac{1}{2}\left(\frac{1}{w}\right)^2 + \frac{3}{2}\left(\frac{1}{w}\right)^2 + w\frac{1}{4!}\left(\frac{1}{w}\right)^4 - \frac{3}{4!}\left(\frac{1}{w}\right)^4 + \cdots \\ &= w - 3 - \frac{1}{2w} + \frac{3}{2w^2} + \frac{1}{4!w^3} - \frac{3}{4!w^4} + \cdots \\ &= z - 1 - \frac{1}{2(z+2)} + \frac{3}{2(z+2)^2} + \frac{1}{4!(z+2)^3} - \frac{3}{4!(z+2)^4} + \cdots \end{aligned}$$

Terms of the form $(z+2)^{-n}$ go on forever in this series, so the point $z = -2$ is an essential singularity. The analytic part of the Laurent expansion is

$$z - 1$$

The principal part of the Laurent expansion is

$$-\frac{1}{2(z+2)} + \frac{3}{2(z+2)^2} + \frac{1}{4!(z+2)^3} - \frac{3}{4!(z+2)^4} + \cdots$$

EXAMPLE 5.15

Given that $f(z) = 1/(e^z - 1) = (1/z) - (1/2) + (1/12)z + \cdots$, describe the nature of any singularities and write down the analytic and principal parts of the expansion. Is the function entire?

SOLUTION

The function is not entire because it has a singularity at $z = 0$. Since this is the only singular point, the function is meromorphic. The principal part of the Laurent expansion includes the single term

$$\frac{1}{z}$$

The analytic part is given by

$$-\frac{1}{2}+\frac{1}{12}z+\cdots$$

EXAMPLE 5.16

What are the singular points of $f(z)=3/(z^2+a^2)$.

SOLUTION

Notice that

$$f(z)=\frac{3}{z^2+a^2}=\frac{3}{(z+ia)(z-ia)}$$

Therefore the function has two isolated singular points at $z=\pm ia$. Since there are a finite number of singular points, the function is meromorphic.

Summary

In this chapter we investigated complex sequences and series. A sequence of complex numbers is a function of the integers. The behavior of a sequence can be investigated in the limit of the argument as it tends to infinity. A sequence can describe an individual term in a series, which can be used to represent a complex function. The convergence of series can be investigated using various tests such as the ratio test. Of particular interest in the study of complex variables, is the Laurent series, and we classify functions of a complex variable by looking at singularities which occur in the series expansion.

Quiz

1. Does the sequence $1+\frac{2z}{n}$ converge? If so find an N so that you can define its limit.
2. Find $\sum_{k=0}^{n}\cos k\theta$.
3. Find the radius of convergence for the Maclaurin expansion of $z\cot z$.
4. Find the radius of convergence for $\sum_{n=1}^{\infty}(3+(-1)^n)^n z^n$.
5. Is the sequence $\left\{\frac{1}{1+nz}\right\}$ convergent? If so over what values of z?

6. Find the Maclaurin expansion of $f(z) = \dfrac{z}{z^4 + 9}$.

7. Describe the convergence of $\sum_{n=1}^{\infty} \dfrac{z^n}{n(n+1)}$.

8. Find the Taylor series expansion of $\sinh z$ about the point $z_0 = \pi i$.

9. Parseval's theorem tells us that if $f(z) = \sum_{n=0}^{\infty} a_n z^n$ then $\dfrac{1}{2\pi}\int_0^{2\pi} \left|f(re^{i\theta})\right|^2 d\theta =$ $\sum_{n=0}^{\infty} |a_n|^2 r^{2n}$. Use it to find a series representation for $\dfrac{1}{2\pi}\int_0^{2\pi} e^{r\cos\theta} d\theta$.

10. Find the Laurent series expansion of $f(z) = \dfrac{1}{z-1} + \dfrac{1}{(z-2)^2}$ for $1 < |z| < 2$.

11. Expand $f(z) = \dfrac{z - \sin z}{z^2}$ in a Laurent series and describe the singularity at $z = 0$.

Complex Integration

The study of elementary calculus involves differentiation and integration. We studied differentiation of complex functions in Chap. 3, now we turn to the problem of integration. It turns out that integration of complex functions is a very elegant procedure. The techniques developed here can not only be used to integrate complex functions but they can also be used as a toolbox to evaluate many integrals of real functions. We start the chapter with a simple evaluation of complex functions that are parameterized by a real parameter t and then introduce contour integration. Complex integration involves integration along a curve.

Complex Functions $w(t)$

Suppose that a complex-valued function $w = f(z)$ is defined in terms of one real variable t as follows:

$$w(t) = u(t) + iv(t) \tag{6.1}$$

and that we are considering an interval $a \le t \le b$.

Now, the definite integral of $w = f(z)$ can be written as

$$\int_a^b w(t)\,dt = \int_a^b u(t)\,dt + i\int_a^b v(t)\,dt$$

The integral of the complex function $w = f(z)$ has been translated into two integrals of the real functions $u(t)$ and $v(t)$. We can integrate these functions using the fundamental theorem of calculus provided that certain conditions are met.

Make the definitions:

$$\frac{dU}{dt} = u(t) \qquad \text{and} \qquad \frac{dV}{dt} = v(t)$$

Then it follows that

$$\begin{aligned}\int_a^b w(t)\,dt &= \int_a^b u(t)\,dt + i\int_a^b v(t)\,dt \\ &= U(b) - U(a) + i\,[V(b) - V(a)]\end{aligned}$$

EXAMPLE 6.1

Compute the integral $\int_0^2 (1-it)^2 dt$.

SOLUTION

The first step is to write the integrand in terms of real and imaginary parts. In this case

$$\begin{aligned}(1-it)^2 &= (1-it)(1-it) \\ &= 1 - i2t - t^2 \\ &= 1 - t^2 - i2t\end{aligned}$$

This leads us to make the following definitions:

$$u(t) = 1 - t^2 \qquad \text{and} \qquad v(t) = -2t$$

The integral can then be written as

$$\begin{aligned}\int_0^2 (1-it)^2 dt &= \int_0^2 u(t)\,dt + \int_0^2 v(t)\,dt \\ &= \int_0^2 (1-t^2)\,dt - i\int_0^2 2t\,dt\end{aligned}$$

These are elementary integrals that are easy to evaluate:

$$\int_0^2 (1-t^2)\,dt - i\int_0^2 2t\,dt = t - \frac{t^3}{3} - it^2 \Big|_0^2 = 2 - \frac{8}{3} - 4i = -\frac{2}{3} - 4i$$

EXAMPLE 6.2

Evaluate $\int_0^{\pi/5} e^{i2t}\,dt$.

SOLUTION

Using tools from elementary calculus we have

$$\int_0^{\pi/4} e^{i2t}\,dt = \frac{1}{2i}e^{i2t}\Big|_0^{\pi/4} = -\frac{i}{2}e^{i2t}\Big|_0^{\pi/4} = -\frac{i}{2}e^{i\pi/2} + \frac{i}{2}$$

Now use Euler's formula:

$$e^{i\pi/2} = \cos(\pi/2) + i\sin(\pi/2) = i$$

And so the integral evaluates to

$$\int_0^{\pi/4} e^{i2t}\,dt = -\frac{i}{2}(i) + \frac{i}{2} = \frac{1+i}{2}$$

Properties of Complex Integrals

If $f(z)$ is a function that depends on one real variable t such that $f = u(t) + iv(t)$ then we can use theorems from the calculus of real variables to handle more complex integrals. Suppose that $\alpha = c + id$ is a complex constant. You will recall from the calculus of real variables that we can pull a constant outside of an integral. The same holds true here, where we have

$$\int_a^b \alpha f\,dt = \int_a^b (c+id)(u+iv)\,dt = (c+id)\int_a^b u\,dt + i(c+id)\int_a^b v\,dt \tag{6.2}$$

Let g be another complex function depending on a single real variable such that $g(t) = r(t) + is(t)$. The integral of the sum or difference $f \pm g$ is

$$\int_a^b (f \pm g)\,dt = \int_a^b f dt \pm \int_a^b g\,dt \tag{6.3}$$

Of course, we can also add the real and imaginary parts of the two functions:

$$\int_a^b (f \pm g)\,dt = \int_a^b (u+iv) \pm (r+is)\,dt = \int_a^b (u \pm r)\,dt + i\int_a^b (v \pm s)\,dt$$

The product of two complex functions of a single real variable can be integrated as follows:

$$\int_a^b (fg)\,dt = \int_a^b (u+iv)(r+is)\,dt = \int_a^b (ur - vs)\,dt + i\int_a^b (vr + us)\,dt \qquad (6.4)$$

As in the calculus of real variables, we can split up an interval $a \le t \le b$. Suppose that $a < c < b$. Then we can write

$$\int_a^b f(t)\,dt = \int_a^c f(t)dt + \int_c^b f(t)\,dt \qquad (6.5)$$

Exchanging the limits of integration introduces a minus sign:

$$\int_a^b f(t)\,dt = -\int_b^a f(t)dt \qquad (6.6)$$

The next example is somewhat contrived, since we could calculate the desired result easily, but it illustrates how the formulas could be applied and gives us practice calculating an integral of a complex function.

EXAMPLE 6.3

Given that $\int_0^{\pi/2} e^{t+it}dt = e^{\pi/2}/2 - [(1-i)/2]$, find $\int_{\pi/4}^{\pi/2} e^{t+it}dt$ by calculating $\int_0^{\pi/4} e^{t+it}dt$.

SOLUTION

The integral is easy to calculate. We have

$$\int_0^{\pi/4} e^{t+it}dt = \int_0^{\pi/4} e^{(1+i)t}dt = \frac{1}{1+i}e^{(1+i)t}\bigg|_0^{\pi/4} = \frac{1}{1+i}(e^{(1+i)\pi/4} - 1)$$

Euler's formula tells us that

$$e^{(1+i)\pi/4} = e^{\pi/4}e^{i\pi/4} = e^{\pi/4}(\cos(\pi/4) + i\sin(\pi/4))$$

A table of trigonometric functions can be consulted to learn that

$$\cos(\pi/4) = \sin(\pi/4) = \frac{\sqrt{2}}{2} = \frac{1}{\sqrt{2}}$$

And so

$$e^{(1+i)\pi/4} = \frac{e^{\pi/4}}{\sqrt{2}}(1+i)$$

Hence the integral is

$$\int_0^{\pi/4} e^{t+it}dt = \frac{1}{1+i}(e^{(1+i)\pi/4} - 1) = \frac{1}{1+i}\left(\frac{e^{\pi/4}}{\sqrt{2}}(1+i) - 1\right) = \frac{e^{\pi/4}}{\sqrt{2}} - \frac{1}{1+i}$$

Writing the last term in standard form we obtain

$$\frac{1}{1+i} = \frac{1}{1+i}\left(\frac{1-i}{1-i}\right) = \frac{1-i}{2}$$

Therefore:

$$\int_0^{\pi/4} e^{t+it}dt = \frac{e^{\pi/4}}{\sqrt{2}} - \frac{1-i}{2}$$

Now we use to write an expression that can be used to find the desired integral:

$$\int_0^{\pi/2} e^{t+it}dt = \int_0^{\pi/4} e^{t+it}dt + \int_{\pi/4}^{\pi/2} e^{t+it}dt$$

$$\Rightarrow \int_{\pi/4}^{\pi/2} e^{t+it}dt = \int_0^{\pi/2} e^{t+it}dt - \int_0^{\pi/4} e^{t+it}dt$$

$$= \frac{e^{\pi/2}}{2} - \left(\frac{1-i}{2}\right) - \left(\frac{e^{\pi/4}}{\sqrt{2}} - \frac{1-i}{2}\right) = \frac{e^{\pi/2}}{2} - \left(\frac{1-i}{2}\right) - \frac{e^{\pi/4}}{\sqrt{2}} + \left(\frac{1-i}{2}\right)$$

$$= \frac{e^{\pi/2}}{2} - \frac{e^{\pi/4}}{\sqrt{2}}$$

Contours in the Complex Plane

So far, we've seen how to evaluate integrals of simple functions of a complex variable—that were defined in terms of a single real parameter we called t. Now it's time to generalize and consider a more general case, where we just say we're integrating a function of a complex variable $f(z)$, where $z \in \mathbb{C}$. This can be done using a technique called *contour integration.*

The reason integrals of complex functions are done the way they are is that while an integral of a real-valued function is defined on an interval of the line, an integral

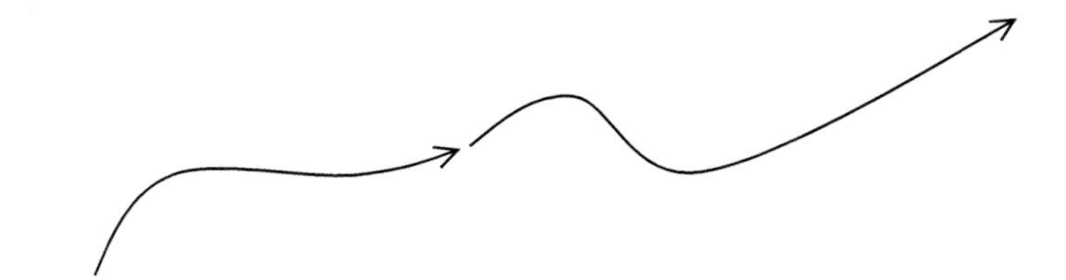

Figure 6.1 A curve $\gamma(t)$ is said to be simple if it does not cross itself.

of a complex-valued function is defined on a curve in the complex plane. We say that a set of points in the complex plane $z = (x, y)$ is an *arc* if $x = x(t)$ and $y = y(t)$ are continuous functions of a real parameter t which ranges over some interval (i.e., $a \le t \le b$). A complex number z can be written as

$$z(t) = x(t) + iy(t)$$

Define a *curve* as a continuous function $\gamma(t)$ that maps a closed interval $a \le t \le b$ to the complex plane. If the curve $\gamma(t)$ that defines a given arc does not cross itself, which means that $\gamma(t_1) \ne \gamma(t_2)$ when $t_1 \ne t_2$, then we say that $\gamma(t)$ is a *simple curve* or *Jordan arc*. A simple curve is illustrated in Fig. 6.1.

If the curve crosses over itself at any point, then it is not simple. An example of this is shown in Fig. 6.2.

The curves in Figs. 6.1 and 6.2 are *open*. If $\gamma(a) = \gamma(b)$, that is $\gamma(t)$ assumes the same value at the endpoints of the interval $a \le t \le b$, but at no other points, then we say that $\gamma(t)$ is a *simple closed curve* or *closed contour*. This is shown in Fig. 6.3.

Formally, we say that a curve $\gamma(t)$ is a simple closed curve if $\gamma(a) = \gamma(b)$ and $\gamma(t)$ is one-to-one.

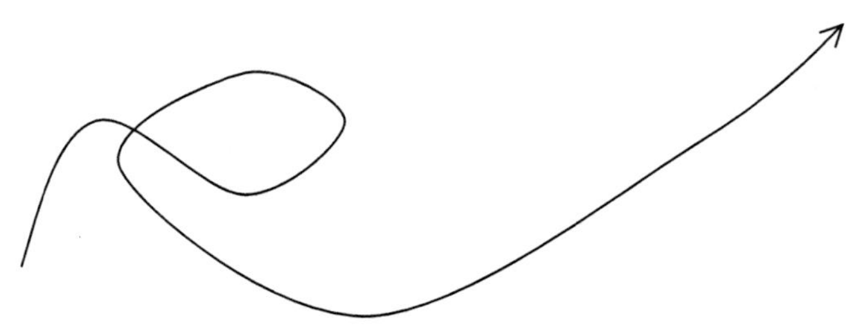

Figure 6.2 A curve which crosses itself at one or more points is not simple.

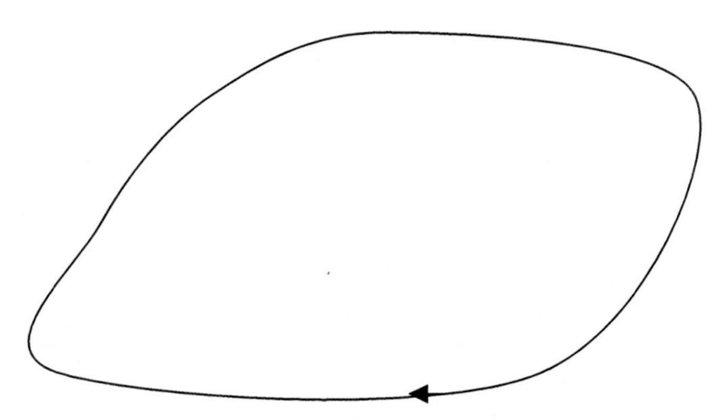

Figure 6.3 A simple, closed curve.

When using contour integration, the sense or direction in which the curve is traversed is important. To understand this, we consider a simple example, the unit circle centered about the origin. For example, consider

$$z = e^{i\theta}$$

where $0 \le \theta \le 2\pi$. If you put in some values as θ ranges over the given interval increasing from 0, you will note that the points sweep out the circle in the *counter-clockwise direction.* To see this, write the points in the complex plane as $z = (x, y)$. Let's plug in a few points:

$$\theta = 0 \Rightarrow z = e^{i0} = \cos 0 + i \sin 0 = (1,0)$$

$$\theta = \pi / 4 \Rightarrow z = e^{i\pi/4} = \cos(\pi / 4) + i \sin(\pi / 4) = \left(\frac{1}{\sqrt{2}}, \frac{1}{\sqrt{2}} \right)$$

$$\theta = \pi / 2 \Rightarrow z = e^{i\pi/2} = \cos(\pi / 2) + i \sin(\pi / 2) = (0,1)$$

$$\theta = \pi \Rightarrow z = e^{i\pi} = \cos \pi + i \sin \pi = (-1,0)$$

Following the curve in the counter-clockwise direction can be said to be in the *positive* sense since it moves with increasing angle. When drawing a contour, we use an arrow to indicate the directional sense we are using to move around it. This is illustrated in Fig. 6.4.

If we move around the curve in the opposite direction, which is clockwise, we'll call that negative because we will be moving opposite to the direction of increasing angles. Now consider the function:

$$z = e^{-i\theta}$$

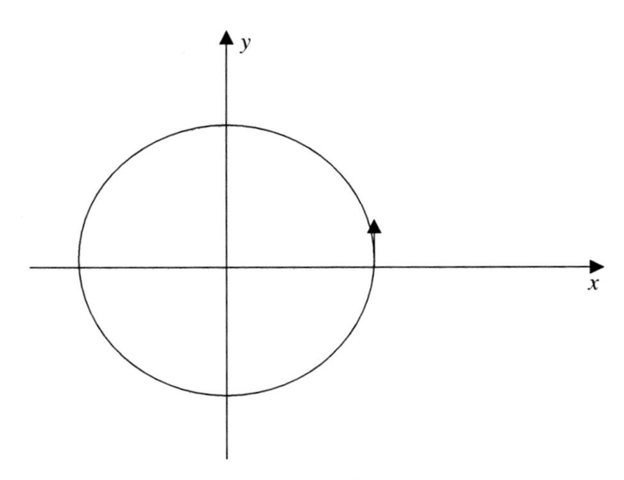

Figure 6.4 A closed contour traversed in the positive sense, which is counter-clockwise. We say that this is in the positive sense because the curve is traversed in the direction of increasing angle θ in the complex plane.

This also describes the unit circle, but we are traversing the circle in the counter-clockwise direction. Notice that

$$\theta = 0 \Rightarrow z = e^{-i0} = \cos 0 - i \sin 0 = (1,0)$$

$$\theta = \pi / 4 \Rightarrow z = e^{-i\pi/4} = \cos(\pi / 4) - i \sin(\pi / 4) = \left(\frac{1}{\sqrt{2}}, -\frac{1}{\sqrt{2}} \right)$$

$$\theta = \pi / 2 \Rightarrow z = e^{-i\pi/2} = \cos(\pi / 2) - i \sin(\pi / 2) = (0,-1)$$

$$\theta = \pi \Rightarrow z = e^{-i\pi} = \cos \pi - i \sin \pi = (-1,0)$$

The case of traversing a circle in the clockwise or negative direction is illustrated in Fig. 6.5.

Complex Line Integrals

In this section, we will formalize what we've done so far with integration a bit. First let's review important properties a function must have so that we can integrate it.

DEFINITION: CONTINUOUSLY DIFFERENTIABLE FUNCTION

Let a function $f(t)$ map the interval $a \le t \le b$ to the real numbers. Formally, we write $f:[a,b] \to \mathbb{R}$. We say that $f(t)$ is continuously differentiable over this interval, which we indicate by writing $f \in C^1([a,b])$ if the following conditions are met:

- The derivative *df/dt* exists on the open interval $a < t < b$.
- The derivative *df/dt* has a continuous extension to $a \le t \le b$.

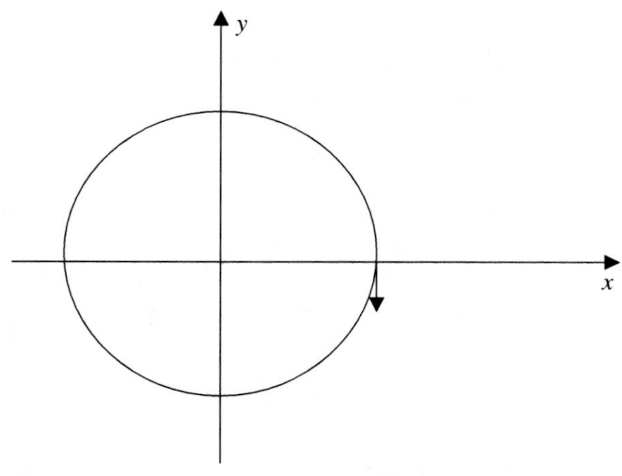

Figure 6.5 Traversing the contour in the negative or clockwise direction.

This allows us to utilize the fundamental theorem of calculus. This tells us that

$$\int_a^b f(t)\,dt = f(b) - f(a) \tag{6.7}$$

Now we will extend this to curves in the complex plane. Suppose that a curve $\gamma(t) = f(t) + ig(t)$.

DEFINITION: CONTINUOUSLY DIFFERENTIABLE CURVE

Let $\gamma(t)$ be a curve, which maps the closed interval $a \le t \le b$ to the complex plane. We say that $\gamma(t)$ is continuous on $a \le t \le b$ if $f(t)$ and $g(t)$ are both continuous on $a \le t \le b$. If $f(t)$ and $g(t)$ are both continuously differentiable functions on $a \le t \le b$, then the curve $\gamma(t)$ is continuously differentiable. This is indicated by writing $\gamma \in C^1([a,b])$.

If $\gamma(t)$ is continuously differentiable and $\gamma(t) = f(t) + ig(t)$, then the derivative is given by

$$\frac{d\gamma}{dt} = \frac{df}{dt} + i\frac{dg}{dt} \tag{6.8}$$

We've already seen that we can write the integral of $w(t) = u(t) + iv(t)$ as $\int_a^b u(t)dt + i\int_a^b v(t)dt$. If the curve $\gamma(t) = f(t) + ig(t)$ is continuously differentiable, then we can write what might be called the fundamental theorem of calculus for curves in the complex plane:

$$\int_a^b \gamma'(t)\,dt = \gamma(b) - \gamma(a) \tag{6.9}$$

This result can be extended further. We consider a domain D in the complex plane and a curve $\gamma(t)$ which maps a real, closed interval $a \le t \le b$ into D. If there is a continuously differentiable function h, which maps D into the real numbers, then the integral along the curve is given by

$$\int_a^b \left(\frac{\partial h}{\partial x}\gamma(t)\frac{df}{dt} + \frac{\partial h}{\partial y}\gamma(t)\frac{dg}{dt} \right) dt = h(\gamma(b)) - h(\gamma(a)) \tag{6.10}$$

DEFINITION: COMPLEX LINE INTEGRAL OR CONTOUR INTEGRAL
Now suppose that the curve $\gamma(t)$ is a simple closed curve. Then the complex line integral of a function $F(z)$ of a complex variable is written as

$$\oint F(z)\,dz = \int_a^b F(\gamma(t))\frac{d\gamma}{dt}\,dt \tag{6.11}$$

The integral in Eq. (6.11) is known as a *contour integral.*

EXAMPLE 6.4
Suppose that $0 \le t \le 1$, $f(z) = z$ and we integrate along the curve $\gamma(t) = 1 + (i-1)t$. Calculate $\int f(z)dz$.

SOLUTION
This can be done by using Eq. (6.11). Given that $\gamma(t) = 1 + (i-1)t$, we see that

$$\frac{d\gamma}{dt} = i - 1$$

We also have that

$$f(z) = f(\gamma(t)) = z = 1 + (i-1)t$$

and so

$$\begin{aligned}
\int_a^b F(\gamma(t))\frac{d\gamma}{dt}\,dt &= \int_0^1 (1 + (i-1)t)(i-1)\,dt \\
&= (i-1)\int_0^1 (1 + (i-1)t)\,dt \\
&= (i-1)\left(t + (i-1)\frac{t^2}{2} \right)\Bigg|_0^1 \\
&= (i-1)\left(\frac{i+1}{2} \right) = -1
\end{aligned}$$

EXAMPLE 6.5

Suppose that $f(z) = z^2 + 1$. Integrate $f(z)$ around the unit circle.

SOLUTION

We can integrate around the unit circle by defining the curve:

$$\gamma(t) = e^{it}$$

The interval mapped by this curve to the complex plane is $0 \le t \le 2\pi$. We find that the derivative of the curve is

$$\frac{d\gamma}{dt} = \frac{d}{dt}(e^{it}) = ie^{it}$$

Using Eq. (6.11) we have

$$\begin{aligned}
\oint f(z)\,dz &= \int_a^b f(\gamma(t))\frac{d\gamma}{dt}dt = \int_0^{2\pi} (\gamma^2 + 1)\frac{d\gamma}{dt}dt \\
&= \int_0^{2\pi} ((e^{it})^2 + 1)(ie^{it})\,dt \\
&= i\int_0^{2\pi} (e^{i3t} + e^{it})\,dt \\
&= \frac{1}{3}e^{i3t} + e^{it}\Big|_0^{2\pi} \\
&= \frac{1}{3}(e^{i6\pi} - 1) + (e^{i2\pi} - 1) = 0
\end{aligned}$$

This result follows since for any even m, $e^{im\pi} = \cos(m\pi) + i\sin(m\pi) = 1 + i0 = 1$.

The Cauchy-Goursat Theorem

Now let's take a turn that we're going to use to develop the groundwork for residue theory, the topic of the next chapter. First let's begin by looking at complex integration once again. We'll dispense with the parameter t and instead focus on functions of x and y. So we have

$$w = f(z) = u(x, y) + iv(x, y)$$

With $z = x + iy$, then

$$dz = dx + idy \tag{6.12}$$

So we can write the integral of a complex function along a curve γ in the following way:

$$\int_\gamma f(z)\,dz = \int_\gamma (u + iv)(dx + idy) = \int_\gamma udx - vdy + i\int_\gamma vdx + udy \tag{6.13}$$

With this in hand, we can define the fundamental theorem of calculus for a function of a complex variable as follows. Suppose that $f(z)$ has an antiderivative. That is:

$$f(z) = \frac{dF}{dz}$$

The fundamental theorem of calculus then becomes

$$\int_\gamma f(z)\,dz = \int_\gamma \frac{dF}{dz}dz = F(z)\Big|_a^b = F(b) - F(a) \tag{6.14}$$

To prove this result, we use $F(z) = U + iV$. We are assuming that *f* and *F* are analytic. Now, using the results of Chap. 3 we know that

$$f(z) = \frac{dF}{dz} = \frac{\partial U}{\partial x} + i\frac{\partial V}{\partial x} = \frac{\partial V}{\partial y} - i\frac{\partial U}{\partial y}$$

and so

$$\begin{aligned}\int_\gamma f(z)dz &= \int_\gamma \frac{dF}{dz}dz \\ &= \int_\gamma \frac{\partial U}{\partial x}dx + \frac{\partial U}{\partial y}dy + i\left(\int_\gamma \frac{\partial V}{\partial x}dx + \frac{\partial V}{\partial y}dy\right)\end{aligned}$$

But since $U = U(x, y)$ using the chain rule we know that

$$dU = \frac{\partial U}{\partial x}dx + \frac{\partial U}{\partial y}dy$$

and similarly for dV. Hence

$$\begin{aligned}\int_\gamma f(z)dz &= \int_\gamma \frac{dF}{dz}dz \\ &= \int_\gamma dU + i\int_\gamma dV \\ &= U\Big|_{z=a}^{z=b} + iV\Big|_a^b \\ &= F(b) - F(a)\end{aligned}$$

The fundamental theorem of calculus allows us to evaluate many integrals in the usual way.

EXAMPLE 6.6

Find $\int_\gamma f(z)dz$ when $f(z) = z^n$ for the case of $z(=a) \neq z(=b)$, and when the curve is closed, that is when $a = b$.

SOLUTION

The fundamental theorem allows us to evaluate the integral in the same way we would in the calculus of real variables. We have

$$\int_\gamma z^n dz = \frac{z^{n+1}}{n+1}\Big|_{z=a}^{z=b}$$

When $z(=a) \neq z(=b)$ this is just

$$\frac{1}{n+1}(b^{n+1} - a^{n+1})$$

If the curve is closed, then $a = b$ and we have the result:

$$\int_\gamma z^n dz = 0$$

This result holds provided that $n \neq -1$.

The case when $n = -1$ introduces us to an interesting phenomena or feature of complex integration. This is the fact that the contour we select for our integration will determine what the result is. First let's do the integral

$$\int_\gamma \frac{dz}{z}$$

using Eq. (6.11), choosing the unit circle as our contour and letting $0 \le t \le 2\pi$. So, $\gamma(t) = e^{it}$. Then we have

$$\int_\gamma \frac{dz}{z} = \int_0^{2\pi} \frac{1}{e^{it}} \frac{d}{dt}(e^{it})dt$$
$$= \int_0^{2\pi} e^{-it}(ie^{it})\,dt$$
$$= i\int_0^{2\pi} dt = 2\pi i$$

Let's look at the integration another way. Now, the domain of $f(z) = 1/z$ is the complex plane less the origin. We write this formally as $\mathbb{C}\setminus\{0\}$. The antiderivative of $f(z)$ is $F(z) = \ln z = \ln r + i\theta$. The domain of the antiderivative is $\mathbb{C}\setminus(-\infty, 0]$. We can do the integral avoiding $\mathbb{C}\setminus(-\infty, 0]$ by taking the contour shown in Fig. 6.6 (notice it is not a closed contour).

To do the integral with this contour, we choose

$$a = re^{-i\pi}, \quad b = re^{i\pi}$$

Note that

$$\ln(b) = \ln(re^{i\pi}) = \ln r + i\pi$$
$$\ln(a) = \ln(re^{-i\pi}) = \ln r - i\pi$$

Using the fundamental theorem of calculus, the integral is

$$\int_\gamma \frac{dz}{z} = \ln(z)\Big|_a^b = \ln(b) - \ln(z) = \ln r + i\pi - (\ln r - i\pi) = i2\pi$$

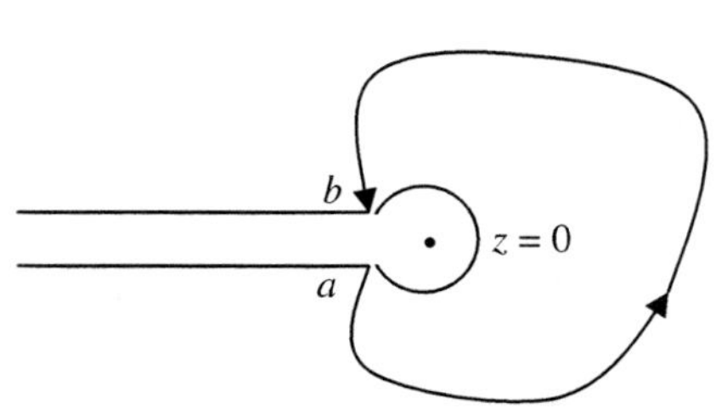

Figure 6.6 A contour that has the point $z = 0$, a singularity of $f(z) = 1/z$, inside the path.

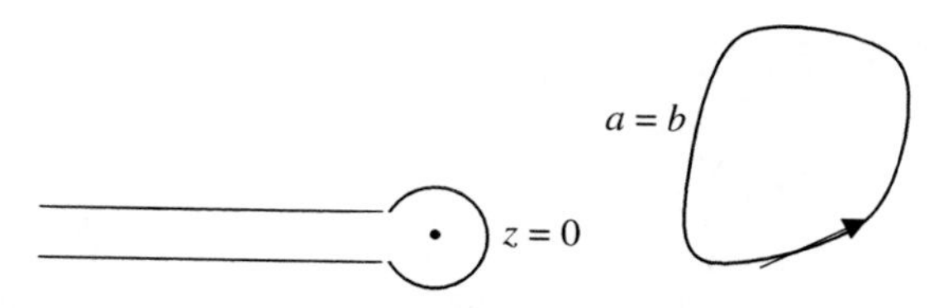

Figure 6.7 We pick a contour that avoids the singularity all together.

This is the same result we obtained using the unit circle and Eq. (6.11). In both cases, the singularity of $f(z)$, the point $z = 0$ was included inside the path. What if we take a closed contour that does *not* include $z = 0$? Such a contour is shown in Fig. 6.7.

This time we have $a = b$ and so

$$\oint \frac{dz}{z} = 0$$

This result suggests the theorem of Cauchy.

THEOREM 6.1: CAUCHY'S INTEGRAL THEOREM

Let U be a simply connected domain and define a function $f : U \to \mathbb{C}$. The Cauchy's integral theorem tells us that if $w = f(z)$ is analytic on a simple, closed curve γ and in its interior, then

$$\oint_\gamma f(z)\,dz = 0 \qquad (6.15)$$

Note that, we take the integration along the curve to be in the positive sense. We can indicate this explicitly by writing

$$\oint_\gamma f(z)dz$$

To prove the theorem, we write

$$\oint_\gamma f(z)\,dz = \oint_\gamma (u + iv)(dx + idy) = \oint_\gamma udx - vdy + i\oint_\gamma vdx + udy$$

We can rewrite this result in terms of partial derivatives and then use Cauchy-Riemann to prove the theorem (we can do this because the assumption of the theorem is that the function is analytic). First we call upon Green's theorem which states that

$$\int_\gamma Pdx + Qdy = \int\int_R \left(\frac{\partial Q}{\partial x} - \frac{\partial P}{\partial y} \right) dxdy$$

where R is a closed region in the plane. Now recall that the Cauchy-Riemann equations tell us

$$\frac{\partial u}{\partial y} = -\frac{\partial v}{\partial x}$$

Green's theorem together with this result gives

$$\oint_{\gamma} udx - vdy = \iint -\frac{\partial u}{\partial y} - \frac{\partial v}{\partial x} dxdy = \iint \frac{\partial v}{\partial x} - \frac{\partial v}{\partial x} dxdy = 0$$

Similarly, we have

$$i\oint_{\gamma} vdx + udy = i\iint \frac{\partial u}{\partial x} - \frac{\partial v}{\partial y} dxdy$$

But the other Cauchy-Riemann equation states that

$$\frac{\partial u}{\partial x} = \frac{\partial v}{\partial y}$$

So the second term vanishes as well. This proves the theorem.

The fundamental theorem of calculus in Eq. (6.14) is actually a consequence of Cauchy's integral theorem. The converse, if you will, of Cauchy's integral theorem is called *Morera's theorem.*

THEOREM 6.2: MORERA'S THEOREM

Let $f(z)$ be a continuous, complex-valued function on an open set D in the complex plane. Suppose that

$$\oint_{\gamma} f(z)\,dz = 0$$

for all closed curves γ. Then it follows that $f(z)$ is analytic.

Next, we extend Cauchy's integral theorem to include singularities in the integrand.

THEOREM 6.3: THE CAUCHY'S INTEGRAL FORMULA

Let $f(z)$ be analytic on a simple closed contour γ and suppose that $f(z)$ is also analytic everywhere on its interior. If the point z_0 is enclosed by γ, then

$$\oint_{\gamma} \frac{f(z)}{z - z_0} dz = 2\pi i\, f(z_0) \tag{6.16}$$

EXAMPLE 6.7

Let γ be the unit circle traversed in a positive sense and suppose that

$$f(z) = \frac{z}{4 - z^2}$$

Find $\oint_\gamma [f(z)/\{z-(i/2)\}]dz$.

SOLUTION

We can apply the Cauchy's integral formula since $z_0 = i/2$ is inside the circle and $f(z)$ is analytic in the given domain (the function has a singularity at $z = \pm 2$, but these points are outside the unit circle). Hence

$$\oint_\gamma \frac{f(z)}{z - i/2} dz = 2\pi i \cdot f(i/2) = 2\pi i \left(\frac{i/2}{4 - (i/2)^2} \right) = -\frac{4\pi}{17}$$

EXAMPLE 6.8

Let $f(z) = 5z - 2$ and γ be the circle defined by $|z| = 2$. Compute

$$\oint_\gamma \frac{5z-2}{z-1} dz$$

SOLUTION

The function $f(z) = 5z - 2$ is clearly analytic on and inside the curve. Also, the point $z = 1$ lies inside $|z| = 2$. So, we can use Eq. (6.16) to evaluate the integral. We have

$$\oint_\gamma \frac{5z-2}{z-1} dz = 2\pi i\, f(z_0) = 2\pi i(5-2) = 6\pi i$$

Summary

In this chapter, complex integration was first considered along a curve parameterized with a single real parameter. Integration in this case is straight forward. We then built up to the Cauchy's integral formula, by developing the fundamental theorem of calculus for a function of a complex variable and then stating and proving Cauchy's integral theorem. In the next chapter, we introduce the elegant theory of residues which is an extension of Cauchy's integral formula.

Quiz

1. Evaluate $\int_0^1 (t-i)^3 dt$.
2. Compute $\int_0^{\pi/6} e^{i2t} dt$.
3. Calculate $\int_0^{\pi/2} e^{t+it} dt$.
4. Find $\int_0^{\pi} e^t \sin t \, dt$ using $\int_0^{\pi} e^{t+it} dt$.
5. Suppose that m and n are integers such that $m \neq n$. Find $\int_0^{2\pi} e^{i(m-n)\theta} d\theta$.
6. Integrate $f(z) = z^2$ around the unit circle which is defined by $\gamma(t) = \cos t + i \sin t$ and $t \in [0, 2\pi)$.
7. Use complex integration to find $\int_0^x \frac{ds}{1+s^2}$.
8. Consider a positively oriented circle with $|z| = 2$. Evaluate $\oint \frac{zdz}{(4-z^2)(z+i)}$.
9. Let γ be the positively oriented unit circle and $f(z) = z$. Evaluate $\oint \frac{zdz}{2z+1}$.
10. Let γ be a positively oriented curve defined by a square with sides located at $x = \pm 3$ and $y = \pm 3$. Evaluate $\oint \frac{\sin z}{\left(z + \frac{\pi}{2}\right)(z^2+16)} dz$.

CHAPTER 7

Residue Theory

In the last chapter, we introduced the notion of complex integration. An important part of our development was the statement of Cauchy's integral formula. In this chapter, we're going to extend this technique using *residue theory.* This is an elegant formulation that not only allows you to calculate many complex integrals, but also gives you a trick you can use to calculate many real integrals. We begin by stating some theorems related to Cauchy's integral formula.

Theorems Related to Cauchy's Integral Formula

We begin the chapter by writing down another form of Cauchy's integral formula. First let's write Eq. (6.16) in the following way:

$$f(a) = \frac{1}{2\pi i} \oint_{\gamma} \frac{f(z)}{z-a} dz \tag{7.1}$$

Now let's take the derivative of this expression, *with respect to a*. This gives

$$f'(a)=\frac{d}{da}\left[\frac{1}{2\pi i}\oint_\gamma \frac{f(z)}{z-a}dz\right]=\frac{1}{2\pi i}\oint_\gamma \frac{d}{da}\left[\frac{f(z)}{z-a}\right]dz=\frac{1}{2\pi i}\oint_\gamma \frac{f(z)}{(z-a)^2}dz$$

We can repeat this process multiple times. That is, take the derivative again. Each time the exponent, which is negative, cancels out the minus sign we pick up by computing the derivative with respect to a of $z-a$. For example, the second derivative is

$$f''(a)=\frac{d}{da}\left(\frac{1}{2\pi i}\oint_\gamma \frac{f(z)}{(z-a)^2}dz\right)=\frac{1}{\pi i}\oint_\gamma \frac{f(z)}{(z-a)^3}dz$$

This process can be continued. For an arbitrary n, we obtain a second Cauchy's integral formula for the nth derivative of $f(a)$:

$$f^{(n)}(a)=\frac{n!}{2\pi i}\oint_\gamma \frac{f(z)}{(z-a)^{n+1}}dz \qquad \text{for } n=1,2,3,... \tag{7.2}$$

There are two facts you should come away with from Cauchy's integral formulas:

- If a function $f(z)$ is known on a simple closed curve γ, then that function is known at all points inside γ. Moreover, all of the functions derivatives can be found inside γ.
- If a function is analytic in a simply connected region of the complex plane, and hence has a first derivative, all of its higher derivatives exist in that simply connected region.

Now we turn to a statement known as *Cauchy's inequality*. This statement is related to Eq. (7.2), which gives us an expression we can use to calculate the derivative of an analytic function in a simply connected region. Consider a circle of radius r, which has the point $z=a$ at its center, and suppose that $f(z)$ is analytic on the circle and inside the circle. Let M be a positive constant such that $|f(z)|\le M$ in the region $|z-a|<r$. Then

$$\left|f^{(n)}(a)\right|\le\frac{Mn!}{r^n} \tag{7.3}$$

The next theorem, which is due to Liouville, tells us that an entire function cannot be bounded unless it is a constant. This statement is called *Liouville's theorem* but it was first proved by Cauchy. So maybe we should call it the Cauchy-Liouville theorem. In any case, it simply says that if $f(z)$ is analytic and bounded in the entire complex plane, that is, $|f(z)|<M$ for some constant M, then $f(z)$ is a constant.

Liouville's theorem implies the *fundamental theorem of algebra.* Consider a polynomial with degree $n \geq 1$ and coefficient $a_n \neq 0$:

$$P(z) = a_0 + a_1 z + a_2 z^2 + \cdots + a_n z^n$$

The fundamental theorem of algebra tells us that every polynomial $P(z)$ has at least one root. The proof follows from Liouville's theorem and the use of a proof by contradiction. Suppose that instead $P(z) \neq 0$ for all z. Then

$$f(z) = \frac{1}{P(z)}$$

is analytic throughout the complex plane and is bounded outside some circle $|z| = r$. Moreover, the assumption that $P(z) \neq 0$ implies that $f = 1/P$ is also bounded for $|z| \leq r$. Hence $1/P(z)$ is bounded in the entire complex plane. Using Liouville's theorem, $1/P(z)$ must be a constant. This is a contradiction, since $P(z) = a_0 + a_1 z + a_2 z^2 + \cdots + a_n z^n$ is clearly not constant. Therefore $P(z)$ must have at least one root such that $P(z) = a_0 + a_1 z + a_2 z^2 + \cdots + a_n z^n = 0$ is satisfied.

Next we state the *maximum modulus theorem* and the *minimum modulus theorem.* The maximum modulus theorem tells us the following. Let $f(z)$ be a complex-valued function which is analytic inside and on a simple closed curve γ. If $f(z)$ is not a constant, then the maximum value of $|f(z)|$ is found on the curve γ.

Now we state the minimum modulus theorem. Assume once again that $f(z)$ is a complex-valued function which is analytic inside and on a simple closed curve γ. If $f(z) \neq 0$ inside γ, then $|f(z)|$ assumes its minimum value on the curve γ.

The next theorem is the *deformation of path theorem.* Consider a domain D in the complex plane, and two curves in D we call γ_1 and γ_2. We suppose that γ_1 is larger than or lies outside of γ_2, and that γ_1 can be deformed into γ_2 without leaving the domain D [that is, we can shrink the first curve down to the second one without crossing any holes or discontinuities in the domain (Fig. 7.1)]. If $f(z)$ is analytic in D then

$$\oint_{\gamma_1} f(z)\,dz = \oint_{\gamma_2} f(z)\,dz \tag{7.4}$$

Next, we state *Gauss' mean value theorem.* Consider a circle γ of radius r centered at the point a. Let $f(z)$ be a function, which is analytic on and inside γ. The mean value of $f(z)$ on γ is given by $f(a)$:

$$f(a) = \frac{1}{2\pi}\int_0^{2\pi} f(a + re^{i\theta})\,d\theta \tag{7.5}$$

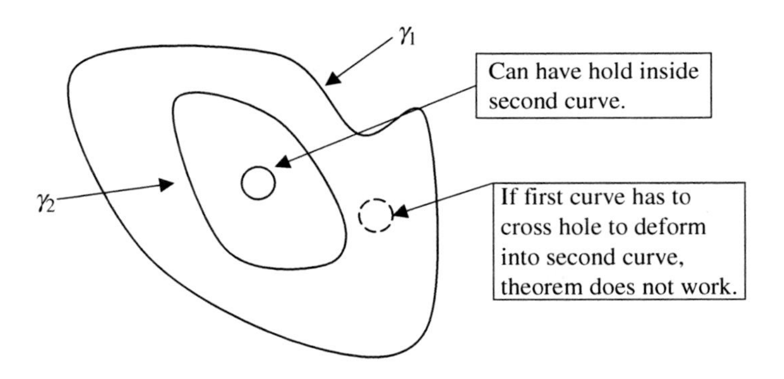

Figure 7.1 A graphic illustration of the deformation of path theorem.

Once again, let $f(z)$ be a function, which is analytic on and inside a simple, closed curve γ. Now assume that $f(z)$ has a finite number of poles inside γ. If M is the number of zeros of $f(z)$ inside γ and N is the number of poles inside γ, the *argument theorem* states that

$$\frac{1}{2\pi i}\oint_{\gamma}\frac{f'(z)}{f(z)}dz = M - N \tag{7.6}$$

Next is a statement of *Rouche's theorem.* Let $f(z)$ and $g(z)$ be two functions, which are analytic inside and on a simple closed curve γ. If $|g(z)| \leq |f(z)|$ on γ, then $f(z)+g(z)$ and $f(z)$ have the same number of zeros inside γ.

Finally, we end our whirlwind tour of theorems and results related to the Cauchy's integral formula with a statement of Poisson's integral formula for a circle. This expresses the value of a harmonic function inside of a circle in terms of its values on the boundary. Let $f(z)$ be analytic inside and on the circle γ, centered at the origin with radius R. Suppose that $z = re^{i\theta}$ is any point inside γ. Then

$$f(z) = \frac{1}{2\pi}\int_0^{2\pi}\frac{(R^2 - r^2)f(\mathrm{Re}^{i\phi})}{R^2 - 2Rr\cos(\theta - \phi) + r^2}d\phi \tag{7.7}$$

EXAMPLE 7.1

This example illustrates the solution of Laplace's equation on a disk. First show that

$$u(r,\theta) = a_0 + \sum_{n=1}^{\infty} a_n r^n \cos n\theta + b_n r^n \sin n\theta$$

is the solution of Laplace's equation on the disc $0 \le r \le 1$ with Dirichlet boundary conditions:

$$\frac{1}{r}\frac{\partial}{\partial r}\left(r\frac{\partial u}{\partial r}\right)+\frac{1}{r^2}\frac{\partial^2 u}{\partial r^2}=0 \qquad 0<r<1, 0\le\theta\le 2\pi$$

$$u(1,\theta)=f(\theta), u(r,\theta) \text{ bounded as } r\to 0$$

Show the coefficients in the series expansion are given by

$$a_0=\frac{1}{2\pi}\int_0^{2\pi} f(\theta)\,d\theta \qquad a_n=\frac{1}{\pi}\int_0^{2\pi} f(\theta)\cos n\theta\,d\theta \qquad b_n=\frac{1}{\pi}\int_0^{2\pi} f(\theta)\sin n\theta\,d\theta$$

Use the result to deduce Poisson's integral formula for a circle of radius one:

$$u(r,\theta)=\frac{1}{2\pi}\int_0^{2\pi}\frac{1-r^2}{1-2r\cos(\theta-\phi)+r^2}f(\phi)\,d\phi$$

SOLUTION

We try separation of variables. Let $u(r,\theta)=R(r)\Theta(\theta)$. Then it follows that

$$\frac{\partial u}{\partial r}=\frac{\partial R}{\partial r}\Theta(\theta) \qquad \frac{\partial^2 u}{\partial r^2}=\frac{\partial^2 R}{\partial r^2}\Theta(\theta) \qquad \frac{\partial^2 u}{\partial \theta^2}=R(r)\frac{\partial^2 \Theta}{\partial \theta^2}$$

The statement of the problem tells us that

$$\frac{\partial^2 u}{\partial r^2}+\frac{1}{r}\frac{\partial u}{\partial r}+\frac{1}{r^2}\frac{\partial^2 u}{\partial \theta^2}=0$$

Hence

$$0=\frac{\partial^2 R}{\partial r^2}\Theta(\theta)+\frac{1}{r}\frac{\partial R}{\partial r}\Theta(\theta)+\frac{1}{r^2}R(r)\frac{\partial^2 \Theta}{\partial \theta^2}$$

We divide every term in this expression by $u(r,\theta)=R(r)\Theta(\theta)$. This allows us to write

$$\frac{r^2}{R}\frac{\partial^2 R}{\partial r^2}+\frac{r}{R}\frac{\partial R}{\partial r}=-\frac{1}{\Theta}\frac{\partial^2 \Theta}{\partial \theta^2}$$

The left-hand side and the right-hand side are functions of r only and θ only, respectively. Therefore they can be equal only if they are both equal to a constant. We call this constant n^2. Then we have the equation in θ:

$$-\frac{1}{\Theta}\frac{d^2\Theta}{d\theta^2} = n^2 \qquad \Rightarrow \frac{d^2\Theta}{d\theta^2} + n^2\Theta = 0$$

Note that partial derivatives can be replaced by ordinary derivatives at this point, since each equation involves one variable only. This familiar differential equation has solution given by

$$\Theta(\theta) = a_n \cos n\theta + b_n \sin n\theta$$

Now, turning to the equation in r, we have

$$\frac{r^2}{R}\frac{d^2R}{dr^2} + \frac{r}{R}\frac{dR}{dr} = n^2 \qquad \Rightarrow r^2\frac{d^2R}{dr^2} + r\frac{dR}{dr} - n^2R = 0$$

You should also be familiar with this equation from the study of ordinary differential equations. It has solution

$$R(r) = c_n r^n + c_{-n} r^{-n}$$

The total solution, by assumption is the product of both solutions, that is, $u(r,\theta) = R(r)\Theta(\theta)$. So we have

$$u(r,\theta) = (c_n r^n + c_{-n} r^{-n})(a_n \cos n\theta + b_n \sin n\theta)$$

The condition that $u(r,\theta)$ is bounded as $r \to 0$ imposes a requirement that the constant $c_{-n} = 0$ since

$$\frac{c_{-n}}{r^n} \to \infty \quad \text{as} \quad r \to 0$$

Therefore, we take $u(r,\theta) = (c_n r^n)(a_n \cos n\theta + b_n \sin n\theta)$. We can just absorb the constant c_n into the other constants, and still designate them by the same letters. Then

$$u(r,\theta) = r^n(a_n \cos n\theta + b_n \sin n\theta)$$

The most general solution is a superposition of such solutions which ranges over all possible values of n. Therefore we write

$$u(r,\theta) = \sum_{n=0}^{\infty} r^n (a_n \cos n\theta + b_n \sin n\theta) = a_0 + \sum_{n=1}^{\infty} r^n (a_n \cos n\theta + b_n \sin n\theta)$$

To proceed, the following orthogonality integrals are useful:

$$\int_0^{2\pi} \sin m\theta \sin n\theta \, d\theta = \begin{cases} \pi\delta_{mn} & \text{for } n \neq 0 \\ 0 & \text{for } n = 0 \end{cases} \tag{7.8}$$

$$\int_0^{2\pi} \cos m\theta \cos n\theta \, d\theta = \begin{cases} \pi\delta_{mn} & \text{for } n \neq 0 \\ 2\pi\delta_{mn} & \text{for } n = 0 \end{cases} \tag{7.9}$$

$$\int_0^{2\pi} \sin m\theta \cos n\theta \, d\theta = 0 \tag{7.10}$$

Here, $\delta_{mn} = 1$ for $m = n, 0$ which is the *Kronecker delta function.* Now we apply the boundary condition $u(1,\theta) = f(\theta)$ for $0 \le \theta \le 2\pi$:

$$f(\theta) = a_0 + \sum_{n=1}^{\infty} r^n (a_n \cos n\theta + b_n \sin n\theta) \tag{7.11}$$

Multiply through this expression by $\sin m\theta$ and integrate. We obtain

$$\begin{aligned} \int_0^{2\pi} f(\theta) \sin m\theta \, d\theta &= a_0 \int_0^{2\pi} \sin m\theta \, d\theta \\ &\quad + \sum_{n=1}^{\infty} \left(\int_0^{2\pi} a_n \cos n\theta \sin m\theta \, d\theta + \int_0^{2\pi} b_n \sin n\theta \sin m\theta \, d\theta \right) \\ &= \sum_{n=1}^{\infty} \left(\int_0^{2\pi} b_n \sin n\theta \sin m\theta \, d\theta \right) = \sum_{n=1}^{\infty} b_n \pi \delta_{mn} = \pi b_m \end{aligned}$$

where Eqs. (7.8)–(7.10) were used. We conclude that

$$b_n = \frac{1}{\pi} \int_0^{2\pi} f(\theta) \sin n\theta \, d\theta$$

Now we return to Eq. (7.11), and multiply by $\cos m\theta$ and integrate. This time

$$\begin{aligned}\int_0^{2\pi} f(\theta)\cos m\theta\, d\theta &= a_0 \int_0^{2\pi} \cos m\theta\, d\theta \\ &\quad + \sum_{n=1}^{\infty}\left(\int_0^{2\pi} a_n \cos n\theta \cos m\theta\, d\theta + \int_0^{2\pi} b_n \sin n\theta \cos m\theta\, d\theta\right) \\ &= \sum_{n=1}^{\infty}\left(\int_0^{2\pi} a_n \sin n\theta \cos m\theta\, d\theta\right) = \sum_{n=1}^{\infty} a_n \pi \delta_{mn} = \pi a_m\end{aligned}$$

Hence

$$a_m = \frac{1}{\pi}\int_0^{2\pi} f(\theta)\cos m\theta\, d\theta$$

To obtain the constant a_0, we integrate without first multiplying by any trig functions, that is:

$$\begin{aligned}\int_0^{2\pi} f(\theta)\, d\theta &= a_0 \int_0^{2\pi} d\theta + \sum_{n=1}^{\infty}\left(\int_0^{2\pi} a_n \cos n\theta\, d\theta + \int_0^{2\pi} b_n \sin n\theta\, d\theta\right) \\ &= a_0 2\pi \\ \Rightarrow a_0 &= \frac{1}{2\pi}\int_0^{2\pi} f(\theta)\, d\theta\end{aligned}$$

This should be obvious since

$$\int_0^{2\pi} \cos n\theta\, d\theta = \frac{1}{n}\sin n\theta\bigg|_0^{2\pi} = 0$$

$$\int_0^{2\pi} \sin n\theta\, d\theta = -\frac{1}{n}\cos n\theta\bigg|_0^{2\pi} = 0$$

Now we are in a position to derive Poisson's formula. We have

$$\begin{aligned}u(r,\theta) &= \frac{1}{2\pi}\int_0^{2\pi} f(\phi)\, d\phi \\ &\quad + \sum_{n=1}^{\infty} r^n \left[\frac{1}{\pi}\left(\int_0^{2\pi} f(\phi)\cos n\phi\, d\phi\right)\cos n\theta + \frac{1}{\pi}\left(\int_0^{2\pi} f(\phi)\sin n\phi\, d\phi\right)\sin n\theta\right]\end{aligned}$$

We can move the summation inside the integrals:

$$u(r,\theta) = \frac{1}{2\pi}\int_0^{2\pi} f(\phi)\, d\phi + \int_0^{2\pi} f(\phi)\frac{1}{\pi}\left(\sum_{n=1}^{\infty} r^n \cos n\theta \cos n\phi\right) d\phi$$

$$+\int_0^{2\pi} f(\phi)\frac{1}{\pi}\left(\sum_{n=1}^{\infty} r^n \sin n\theta \sin n\phi\right) d\phi$$

$$= \frac{1}{2\pi}\int_0^{2\pi} d\phi f(\phi)\left\{1 + 2\sum_{n=1}^{\infty} r^n \cos n\theta \cos n\phi + 2\sum_{n=1}^{\infty} r^n \sin n\theta \sin n\phi\right\}$$

$$= \frac{1}{2\pi}\int_0^{2\pi} d\phi f(\phi)\left\{1 + 2\sum_{n=1}^{\infty} r^n (\cos n\theta \cos n\phi + \sin n\theta \sin n\phi)\right\}$$

Now recall that

$$\cos n\theta \cos n\phi + \sin n\theta \sin n\phi = \cos(n(\theta - \phi))$$

It's also true that

$$1 - 2\sum_{n=1}^{\infty} r^n \cos[n(\theta - \phi)] = \frac{1 - r^2}{1 - 2r\cos(\theta - \phi) + r^2} \qquad (7.12)$$

So, we arrive at the Poisson formula for a disc of radius one:

$$u(r,\theta) = \frac{1}{2\pi}\int_0^{2\pi} \frac{1 - r^2}{1 - 2r\cos(\theta - \phi) + r^2} f(\phi)\, d\phi$$

This tells us that the value of a harmonic function at a point inside the circle is the average of the boundary values of the circle.

The Cauchy's Integral Formula as a Sampling Function

The *Dirac delta function* has two important properties. First if we integrate over the entire real line then the result is unity:

$$\int_{-\infty}^{\infty} \delta(x)\, dx = 1$$

Second, it acts as a *sampling* function—that is, it picks out the value of a real function $f(x)$ at a point:

$$\int_{-\infty}^{\infty} f(x)\,\delta(x-a)dx = f(a)$$

In complex analysis, the function $1/z$ plays an analogous role. It has a singularity at $z = 0$, and

$$\frac{1}{2\pi i}\oint_{\gamma}\frac{dz}{z} = \begin{cases} 0 & \text{if 0 is not in the interior of } \gamma \\ 1 & \text{if 0 is in the interior of } \gamma \end{cases}$$

It also acts as a sampling function *for analytic functions* $f(z)$ in that

$$f(a) = \frac{1}{2\pi i}\oint_{\gamma}\frac{f(z)dz}{z-a}$$

Some Properties of Analytic Functions

Now we are going to lay some more groundwork before we state the residue theorem. In this section, we consider some properties of analytic functions.

AN ANALYTIC FUNCTION HAS A LOCAL POWER SERIES EXPANSION

Suppose that a function $f(z)$ is analytic inside a disc centered at a point a of radius r: $|z-a| < r$. Then $f(z)$ has a power series expansion given by

$$f(z) = \sum_{n=0}^{\infty} a_n (z-a)^n \tag{7.13}$$

The coefficients of the expansion can be calculated using the Cauchy's integral formula in Eq. (7.2):

$$a_n = \frac{f^{(n)}(a)}{n!} \tag{7.14}$$

INTEGRATION OF THE POWER SERIES EXPANSION GIVES ZERO

Note the following result:

$$\int_{\gamma} (z-a)^m \, dz = \begin{cases} 0 & \text{if } m \neq -1 \\ \ln(z-a) & \text{if } m = -1 \end{cases}$$

Hence

$$\int_{\gamma} f(z)\, dz = \sum_{n=0}^{\infty} a_n \int_{\gamma} (z-a)^n \, dz$$

since n is never equal to -1.

A FUNCTION *f*(*z*) THAT IS ANALYTIC IN A PUNCTURED DISC HAS A LAURENT EXPANSION

Consider the punctured disc of radius r centered at the point a. We denote this by writing $0 < |z-a| < r$. If $f(z)$ is analytic in this region, it is analytic inside the disc but not at the point a. In this case, the function has a Laurent expansion:

$$f(z) = \sum_{n=-\infty}^{\infty} a_n (z-a)^n \tag{7.15}$$

As stated in Chap. 5, we can classify the points at which the function blows up or goes to zero. A *removable singularity* is a point a at which the function appears to be undefined, but it can be shown by writing down the Laurent expansion that in fact the function is analytic at a. In this case the Laurent expansion in Eq. (7.15) assumes the form

$$f(z) = \sum_{n=k}^{\infty} a_n (z-a)^n$$

where $k \geq 0$. Then it turns out the point $z = a$ is a *zero* of order k.

On the other hand, suppose that the series expansion retains terms with $n < 0$:

$$f(z) = \sum_{n=-k}^{\infty} a_n (z-a)^n$$

Then we say that the point $z = a$ is a *pole* of order k. Simply put, a pole is a point that behaves like the point $z = 0$ for $g(z) = 1/2$. That is, as $z \to a$, then $f(z) \to \infty$

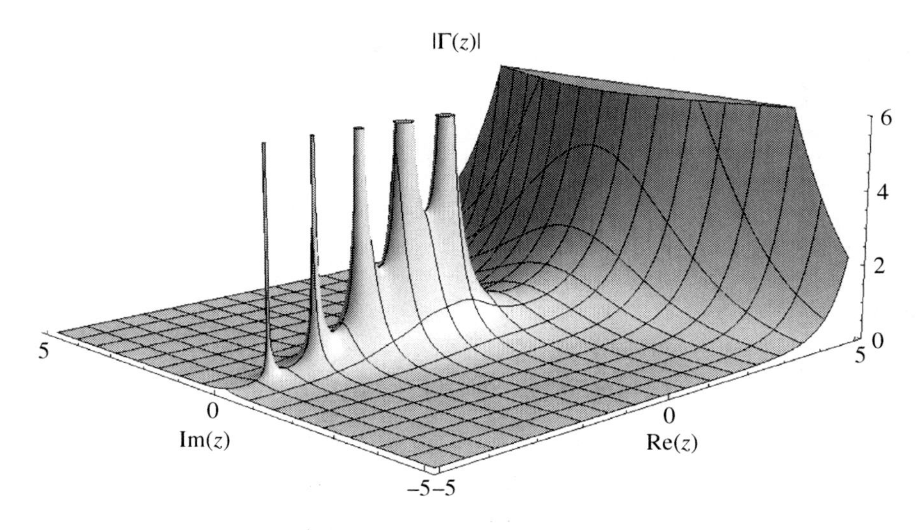

Figure 7.2 When the real part of z is negative, the modulus of the gamma function blows to infinity at several points. These points are the poles of the function.

A function might have multiple poles. For example, in Fig. 7.2 we illustrate the poles of the modulus of the gamma function, $|\Gamma(z)|$, which are points where the function blows up.

A Laurent series expansion of this type can be split into two parts:

$$f(z) = \sum_{n=-k}^{\infty} a_n (z-a)^n = \sum_{n=-k}^{-1} a_n (z-a)^n + \sum_{n=0}^{\infty} a_n (z-a)^n = F + G$$

The second series, which we have denoted by G, looks like a plain old Taylor expansion. The other series, which we have denoted by F, is called the *principal part* and it includes the singularities (the real ones—the poles) of the function.

EXAMPLE 7.2

Is the point $z = 0$ a removable singularity of $f(z) = (\sin z)/z$?

SOLUTION

At first glance, the behavior of the function at $z = 0$ can't really be determined. To see what's going on we expand the sin function in Taylor:

$$f(z) = \frac{\sin z}{z} = \frac{1}{z}\left(z - \frac{1}{3!}z^3 + \frac{1}{5!}z^5 + \cdots \right)$$

$$= 1 - \frac{1}{3!}z^2 + \frac{1}{5!}z^4 + \cdots$$

From this expression, it's easy to see that

$$\lim_{z\to 0} f(z) = \lim_{z\to 0} \frac{\sin z}{z} = \lim_{z\to 0} 1 - \frac{1}{3!}z^2 + \frac{1}{5!}z^4 + \cdots = 1$$

Therefore, the point $z = 0$ is a zero of order one.

EXAMPLE 7.3

Describe the nature of the singularities of $f(z) = e^z/z$.

SOLUTION

We follow the same procedure used in Example 7.2. First expand in Taylor:

$$f(z) = \frac{e^z}{z} = \frac{1}{z}\left(1 + z + \frac{z^2}{2!} + \frac{z^3}{3!} + \cdots\right) = \frac{1}{z} + 1 + \frac{z}{2} + \frac{z^2}{6} + \cdots$$

The principal part of this series expansion is given by $1/z$. It follows that the point $z = 0$ is a pole of order one.

EXAMPLE 7.4

Is the point $z = 0$ a removable singularity of $f(z) = (\sin z)/z^4$?

SOLUTION

Contrast this solution with that found in Example 7.2. Expanding in Taylor we find

$$f(z) = \frac{\sin z}{z^4} = \frac{1}{z^4}\left(z - \frac{1}{3!}z^3 + \frac{1}{5!}z^5 - \frac{1}{7!}z^7 - \cdots\right)$$
$$= \frac{1}{z^3} - \frac{1}{6z} + \frac{1}{5!}z - \frac{1}{7!}z^3 + \cdots$$

This time, the singularity cannot be removed. So the point $z = 0$ is a pole. The principal part in this series expansion is

$$\frac{1}{z^3} - \frac{1}{6z}$$

The leading power (most negative power) in the expansion gives the order of the pole. Hence $z = 0$ is a pole of order three.

ESSENTIAL SINGULARITY

Next we consider the *essential singularity.* In this case, the Laurent series expansion of the function includes a principal part that is nonterminating. That is, all terms out to minus infinity are included in the Laurent expansion with negative n, that is, there are no nonzero terms in the expansion for $n < 0$:

$$f(z) = \sum_{n=-\infty}^{\infty} a_n (z-a)^n$$

EXAMPLE 7.5

Describe the nature of the singularity at $z = 0$ for $f(z) = e^{1/z}$.

SOLUTION

This function is the classic example used to illustrate an essential singularity. We just write down the series expansion:

$$\begin{aligned} f(z) &= e^{\frac{1}{z}} \\ &= 1 + \frac{1}{z} + \frac{1}{2}\left(\frac{1}{z}\right)^2 + \frac{1}{6}\left(\frac{1}{z}\right)^3 + \frac{1}{4!}\left(\frac{1}{z}\right)^4 + \frac{1}{5!}\left(\frac{1}{z}\right)^5 + \cdots \\ &= 1 + z^{-1} + \frac{1}{2}z^{-2} + \frac{1}{6}z^{-3} + \frac{1}{4!}z^{-4} + \frac{1}{5!}z^{-5} + \cdots \end{aligned}$$

This series expansion has a nonterminating principal part. Therefore $z = 0$ is an essential singularity.

The Residue Theorem

Now we're in a position where we can describe one of the central results of complex analysis, the *residue theorem.* We consider a function $f(z)$ in a region enclosed by a curve γ that includes isolated singularities at the points $z_1, z_2, \ldots, z_k$. The function is analytic everywhere on the curve and inside it except at the singularities. This is illustrated in Fig. 7.3.

We can use the deformation of path theorem to shrink the curve down. In fact, we can shrink it down into isolated curves enclosing each singularity. This is shown in Fig. 7.4.

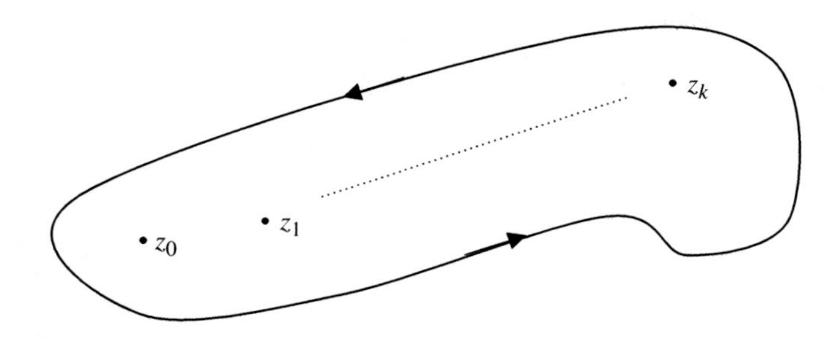

Figure 7.3 A function $f(z)$ is analytic in a certain region enclosed by a curve, except at a set of isolated singularities.

After application of the deformation of path theorem, the integral is broken up into a sum of integrals about each singular point:

$$\int_\gamma f(z)\,dz = \sum_{j=1}^{k} \int_{\gamma_j} f(z)\,dz$$

This expression can be written in terms of the Laurent expansion. Note that there will be a series expansion (which is local) about each singular point:

$$\int_{\gamma_j} f(z)\,dz = \int_{\gamma_j} \sum_{n=-\infty}^{\infty} a_n^j (z-z_j)^n\,dz = \sum_{n=-\infty}^{\infty} a_n^j \int_{\gamma_j} (z-z_j)^n\,dz = a_{-1}^j 2\pi i$$

We call the coefficient in the expansion a_{-1}^j the *residue*. Summing over all of the integrals for each singular point, we get the residue theorem. This states that the integral is proportional to the sum of the residues:

$$\oint_\gamma f(z)\,dz = 2\pi i \sum_{j=1}^{k} \text{residues} \tag{7.16}$$

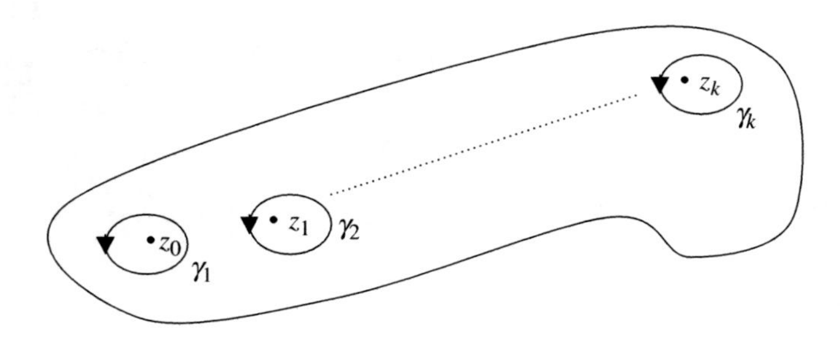

Figure 7.4 If the region is simply connected, we can apply the deformation of path theorem to shrink the curve down, until we have circles around each isolated singular point.

Residues are computed by finding the limit of the function $f(z)$ as z approaches each singularity. This is done for a singularity at $z = a$ as follows:

$$\text{residue} = \lim_{z \to a} \frac{1}{(k-1)!} \frac{d^{k-1}}{dz^{k-1}} \left[(z-a)^k f(z) \right] \tag{7.17}$$

where k is the order of the singularity.

EXAMPLE 7.6

Compute the integral

$$\oint_\gamma \frac{5z-2}{z(z-2)} dz$$

where γ is a circle of radius $r = 3$ centered at the origin.

SOLUTION

The singularities of this function are readily identified to be located at $z = 0, 2$. Both singularities are enclosed by the curve, since $|z| < 3$ in both cases. To find each residue, we compute the limit of the function for each singularity. The residue corresponding to $z = 0$ is

$$\lim_{z \to 0} z \frac{5z-2}{z(z-2)} = \lim_{z \to 0} \frac{5z-2}{(z-2)} = \frac{-2}{-2} = 1$$

The residue corresponding to the singularity at $z = 2$ is

$$\lim_{z \to 2} (z-2) \frac{5z-2}{z(z-2)} = \lim_{z \to 2} \frac{5z-2}{z} = \frac{8}{2} = 4$$

Therefore using Eq. (7.16) the integral evaluates to

$$\oint_\gamma \frac{5z-2}{z(z-2)} dz = 2\pi i \sum residues = 2\pi i(1+4) = 10\pi i$$

EXAMPLE 7.7

Compute the integral of $\int_\gamma [(\cosh z)/z^3] dz$, where γ is the unit circle centered about the origin.

SOLUTION

The function has a singularity at $z = 0$ of 3d order. Using

$$f^{(n)}(a) = \frac{n!}{2\pi i} \oint_\gamma \frac{f(z)}{(z-a)^{n+1}} dz$$

We have

$$\oint \frac{\cosh z}{z^3} dz = \frac{2\pi i}{2!} \frac{d^2}{dz^2} (\cosh z)\Big|_{z=0} = \pi i (\cosh z)\Big|_{z=0} = \pi i$$

Evaluation of Real, Definite Integrals

One of the most powerful applications of the residue theorem is in the evaluation of definite integrals of functions of a real variable. We start by considering

$$\int_0^{2\pi} f(\cos\theta, \sin\theta)\, d\theta$$

Now, write the complex variable z in polar form on the unit circle, that is, let $z = e^{i\theta}$. Notice that

$$dz = ie^{i\theta} d\theta \qquad \Rightarrow d\theta = \frac{1}{iz} dz = -\frac{i}{z} dz$$

As θ increases from 0 to 2π, one sees that the complex variable z moves around the unit circle in a counter clockwise direction. Using Euler's formula, we can also rewrite $\cos\theta$ and $\sin\theta$ in terms of complex variables. In the first case:

$$\cos\theta = \frac{e^{i\theta} + e^{-i\theta}}{2} = e^{-i\theta}\left(\frac{e^{i2\theta} + 1}{2}\right) = \frac{e^{i2\theta} + 1}{2e^{i\theta}} = \frac{z^2 + 1}{2z}$$

Similarly, we find that

$$\sin\theta = \frac{z^2 - 1}{2iz}$$

Taking these facts together, we see that $\int_0^{2\pi} f(\cos\theta, \sin\theta)\, d\theta$ can be rewritten as a contour integral in the complex plane. We only need to include residues that are inside the unit circle.

EXAMPLE 7.8

Compute $\int_0^{2\pi} [d\theta/(24 - 8\cos\theta)]$.

SOLUTION

Using $d\theta = -(i/z)dz$ together with $\cos\theta = (z^2+1)/2z$ we have

$$\int_0^{2\pi} \frac{d\theta}{24-8\cos\theta} = -i\oint \frac{dz}{z\left[24-8\left(\frac{z^2+1}{2z}\right)\right]}$$

$$= -i\oint \frac{dz}{24z-4z^2-4}$$

$$= i\oint \frac{dz}{4z^2-24z+4}$$

$$= \frac{i}{4}\oint \frac{dz}{z^2-6z+1}$$

We will choose the unit circle for our contour. To find the singularities, we find the roots of the denominator. Some algebra shows that they are located at $z = 3 \pm 2\sqrt{2}$.

The first residue is given by

$$\lim_{z\to 3+2\sqrt{2}} \left(z-3-2\sqrt{2}\right)\frac{1}{\left(z-3-2\sqrt{2}\right)\left(z-3+2\sqrt{2}\right)} = \frac{1}{4\sqrt{2}}$$

The residue corresponding to $z = 3-2\sqrt{2}$ is given by

$$\lim_{z\to 3-2\sqrt{2}} \left(z-3+2\sqrt{2}\right)\frac{1}{\left(z-3-2\sqrt{2}\right)\left(z-3+2\sqrt{2}\right)} = -\frac{1}{4\sqrt{2}}$$

You should always check that your singularities lie inside the curve you are using to integrate. If they do not, they do not contribute to the integral. In this case, both residues do not contribute. This is because

$$z = 3+2\sqrt{2} > 1$$

lies outside the unit circle. So we will only include the second residue, because the singularity it corresponds to, $z = 3-2\sqrt{2} < 1$ and so is inside the unit circle.

Using Eq. (7.16) we have

$$\oint \frac{dz}{z^2-6z+1} = 2\pi i \sum \text{residues} = 2\pi i\left(-\frac{1}{4\sqrt{2}}\right) = -\frac{\pi i}{2\sqrt{2}}$$

Hence

$$\int_0^{2\pi} \frac{d\theta}{24-8\cos\theta} = \frac{i}{4}\oint \frac{dz}{z^2-6z+1} = \frac{i}{4}\left(-\frac{\pi i}{2\sqrt{2}}\right) = \frac{\pi}{8\sqrt{2}}$$

The next type of definite integral we consider is one of the form

$$\int_{-\infty}^{\infty} f(x)\begin{Bmatrix}\cos mx \\ \sin mx\end{Bmatrix} dx$$

This type of integral can be converted into a contour integral of the form

$$\oint f(z)\, e^{imz}\, dz \tag{7.18}$$

To obtain the desired result, we take the real or imaginary part of Eq. (7.18) depending on whether or not a cos or sin function is found in the original integral. A useful tool when evaluating integrals of the form in Eq. (7.18) is called *Jordan's lemma.* Imagine that we choose γ to be a semicircle located at the origin and in the upper half plane, as illustrated in Fig. 7.5.

Jordan's lemma states that

$$\lim_{R\to\infty} \int_{C_1} f(z)\, e^{mz} dz = 0 \tag{7.19}$$

Jordan's lemma does not hold in all cases. To use Eq. (7.19), if $m > 0$ then it must be the case that $|f(z)| \to 0$ as $R \to \infty$. We can also apply it in the following case:

$$\lim_{R\to\infty} \int_{C_1} f(z)\, dz = 0$$

provided that $|f(z)| \to 0$ faster than $1/z$ as $R \to \infty$.

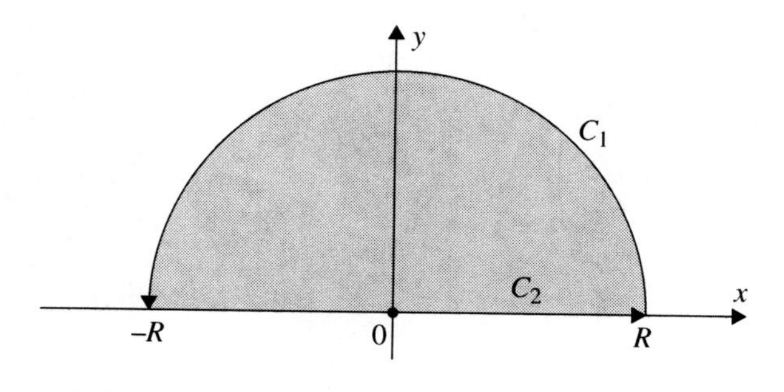

Figure 7.5 A semicircle in the upper half plane, of radius R.

EXAMPLE 7.9

Compute $\int_{-\infty}^{\infty}[(\cos kx)/x^2]dx$.

SOLUTION

We can compute this integral by computing

$$\int_{-\infty}^{\infty}\frac{\cos kx}{x^2}dx = \text{Re}\, I_z$$

where

$$I_z = P\int_{-\infty}^{\infty}\frac{e^{ikz}}{z^2}dz$$

The P stands for *principal part*. To do the integral, we will take a circular contour in the upper half plane which omits the origin. This is illustrated in Fig. 7.6.

Now we can write out the integral piecewise, taking little chunks along the curves C_1 and C_2. Note that when directly on the real axis, we set $z \to x$. This gives

$$\oint\frac{e^{ikz}}{z^2}dz = \int_{-R}^{r}\frac{e^{ikx}}{x^2}dx + \int_{C_2}\frac{e^{ikz}}{z^2}dz + \int_{r}^{R}\frac{e^{ikx}}{x^2}dx + \int_{C_1}\frac{e^{ikz}}{z^2}dz = 0$$

The entire sum of these integrals equals zero because the contour encloses no singular points. However, individual integrals in this expression are not all zero. By Jordan's lemma:

$$\int_{C_1}\frac{e^{ikz}}{z^2}dz = 0$$

So, we only need to calculate the residues for the curve C_2. This curve is in the *clockwise direction*, so we need to add a minus sign when we do our calculation.

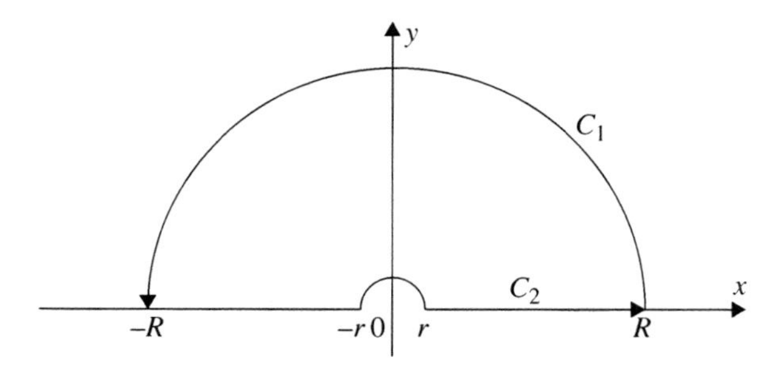

Figure 7.6 We use a semicircular contour in the upper half plane, omitting the origin using a small semicircle or radius r that gives us a curve that omits the origin.

Also, up to this point, we have been using full circles in our calculations. The curve in this case is a semicircle, so Eq. (7.16) is written as

$$\oint_{\gamma} f(z)\,dz = -\pi i \sum_{j=1}^{k} \text{residues}$$

The singularity at $z = 0$ is inside the curve C_2, of order two. The residue corresponding to this singularity is

$$\frac{d}{dz}\left(z^2 \frac{e^{ikz}}{z^2}\right)_{z=0} = ike^{ikz}\Big|_{z=0} = ik$$

Therefore

$$\int_{C_2} \frac{e^{ikz}}{z^2}\,dz = -i\pi(ik) = \pi k$$

Now

$$\oint \frac{e^{ikz}}{z^2}\,dz = \int_{-R}^{r} \frac{e^{ikx}}{x^2}\,dx + \int_{r}^{R} \frac{e^{ikx}}{x^2}\,dx + \int_{C_1} \frac{e^{ikz}}{z^2}\,dz = 0$$

$$P\int_{-\infty}^{\infty} \frac{e^{ikx}}{x^2}\,dx = \int_{-R}^{r} \frac{e^{ikx}}{x^2}\,dx + \int_{r}^{R} \frac{e^{ikx}}{x^2}\,dx \qquad \text{as } r \to 0 \qquad R \to \infty \Rightarrow$$

$$P\int_{-\infty}^{\infty} \frac{e^{ikx}}{x^2}\,dx + \int_{C_1} \frac{e^{ikz}}{z^2}\,dz = 0$$

Therefore we find that

$$\int_{-\infty}^{\infty} \frac{\cos kx}{x^2}\,dx = \operatorname{Re} I_z = P\int_{-\infty}^{\infty} \frac{e^{ikx}}{x^2}\,dx = -\int_{C_1} \frac{e^{ikz}}{z^2}\,dz = -\pi k$$

Integral of a Rational Function

The integral of a rational function $f(x)$

$$\int_{-\infty}^{\infty} f(x)\,dx$$

can be calculated by computing

$$\oint f(z)\,dz$$

using the contour shown in Fig. 7.5, which consists of a line along the x axis from $-R$ to R and a semicircle above the x axis the same radius. Then we take the limit $R \to \infty$.

EXAMPLE 7.10

Consider the Poisson kernel

$$p_y(x) = \frac{1}{\pi} \frac{y}{x^2 + y^2}$$

Treating y as a constant, use the residue theorem to show that its Fourier transform is given by $[1/(2\pi)]e^{-|k|y}$.

SOLUTION

The Fourier transform of a function $f(x)$ is given by the integral

$$F(k) = \frac{1}{2\pi} \int_{-\infty}^{\infty} f(x)\, e^{-ikx} dx \tag{7.20}$$

So, we are being asked to evaluate the integral

$$I = \frac{1}{2\pi} \int_{-\infty}^{\infty} \frac{1}{\pi} \frac{y}{x^2 + y^2} e^{-ikx} dx$$

We do this by considering the contour integral

$$\oint \frac{1}{2\pi^2} \frac{y}{z^2 + y^2} e^{-ikz} dz$$

First, note that

$$\frac{1}{2\pi^2} \frac{y}{x^2 + y^2} = \frac{1}{2\pi^2} \frac{y}{(z + iy)(z - iy)}$$

Therefore, there are two simple poles located at $z = \pm iy$. These lie directly on the y axis, one in the upper half plane and one in the lower half plane. To get the right answer for the integral we seek, we need to compute using both cases. First we consider the pole in the upper half plane. The residue corresponding to $z = +iy$ is

$$a_{-1} = (z - iy) \frac{ye^{-ikz}}{2\pi^2 (z + iy)(z - iy)} \bigg|_{z=+iy} = \frac{ye^{ky}}{2\pi^2 (2iy)} = \frac{1}{4\pi^2 i} e^{ky}$$

Applying the residue theorem, we find that

$$I = \frac{1}{2\pi}\int_{-\infty}^{\infty}\frac{1}{\pi}\frac{y}{x^2+y^2}e^{-ikx}dx = 2\pi i\left(\frac{1}{4\pi^2 i}e^{ky}\right) = \frac{e^{ky}}{2\pi}$$

However, now let's consider enclosing the other singularity, which would be an equally valid approach. The singularity is located at $z = -iy$, which is below the x axis, so we would need to use a semicircle in the lower half plane to enclose it. This time the residue is

$$a_{-1} = (z+iy)\frac{ye^{-ikz}}{2\pi^2(z+iy)(z-iy)}\Big|_{z=-iy} = -\frac{1}{4\pi^2 i}e^{-ky}$$

Using this result, we obtain

$$I = \frac{1}{2\pi}\int_{-\infty}^{\infty}\frac{1}{\pi}\frac{y}{x^2+y^2}e^{-ikx}dx = 2\pi i\left(\frac{1}{4\pi^2 i}e^{-ky}\right) = \frac{e^{-ky}}{2\pi}$$

Combining both results gives the correct answer, which is

$$I = \frac{e^{-|k|y}}{2\pi}$$

EXAMPLE 7.11

Compute the integral given by

$$I = \int_{-\infty}^{\infty}\frac{x^2}{1+x^4}\,dx$$

SOLUTION

This integral is given by

$$I = 2\pi i\sum \text{residues in upper half plane}$$

We find the residues by considering the complex function

$$f(z) = \frac{z^2}{z^4+1}$$

The singularities are found by solving the equation

$$z^4 + 1 = 0$$

This equation is solved by $z = (-1)^{1/4}$. But, remember that $-1 = e^{i\pi}$.

That is, there are four roots given by

$$z = e^{1/4(i\pi)(2n+1)} = \begin{cases} e^{i\pi/4} \\ e^{i3\pi/4} \\ e^{i5\pi/4} \\ e^{i7\pi/4} \end{cases}$$

These are shown in Fig. 7.7. Notice that two of the roots are in the upper half plane, while two of the roots are in the lower half plane. We *reject* the roots in the lower half plane because we are choosing a closed semicircle in the upper half plane (as in Fig. 7.5) as our contour. We only consider the singularities that are inside the contour, the others do not contribute to the integral.

We proceed to compute the two residues. They are all simple so in the first case we have

$$\begin{aligned}
&\lim_{z\to e^{i\pi/4}} \frac{(z-e^{i\pi/4})z^2}{(z-e^{i\pi/4})(z-e^{i3\pi/4})(z-e^{i5\pi/4})(z-e^{i7\pi/4})} \\
&= \lim_{z\to e^{i\pi/4}} \frac{z^2}{(z-e^{i3\pi/4})(z-e^{i5\pi/4})(z-e^{i7\pi/4})} \\
&= \frac{i}{(e^{i\pi/4}-e^{i3\pi/4})(e^{i\pi/4}-e^{i5\pi/4})(e^{i\pi/4}-e^{i7\pi/4})} \\
&= \frac{i}{2e^{i\pi/4}(e^{i\pi/2}-e^{i3\pi/2})} = -\frac{1}{4}e^{i3\pi/4}
\end{aligned}$$

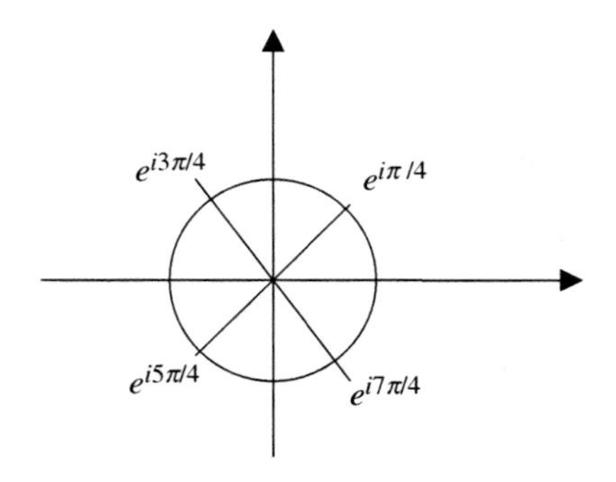

Figure 7.7 An illustration of the roots in Example 7.11. For our contour, we will enclose the upper half plane, so we ignore the roots that lie in the lower half plane.

And so, shows that the residue corresponding to the pole at $z = \exp(i3\pi/4)$ is given by $z = e^{i3\pi/4} \Rightarrow -(1/4)e^{i\pi/4}$. Hence

$$\begin{aligned}\sum \text{residues} &= -\frac{1}{4}e^{i\pi/4} - \frac{1}{4}e^{i3\pi/4} \\ &= -\frac{1}{4}(\cos\pi/4 + i\sin\pi/4 + \cos 3\pi/4 + i\sin 3\pi/4) \\ &= -\frac{1}{4}\frac{1}{\sqrt{2}}(1+i-1+i) = -\frac{1}{4\sqrt{2}}2i = -\frac{i}{2\sqrt{2}}\end{aligned}$$

Therefore the integral evaluates to

$$I = 2\pi i \sum \text{residues in upper half plane} = 2\pi i\left(-\frac{i}{2\sqrt{2}}\right) = \frac{\pi}{\sqrt{2}}$$

EXAMPLE 7.12

Compute $\int_{-\infty}^{\infty} [(\cos x)/(x^2 - 2x + 2)]dx$.

SOLUTION

We can compute this integral by considering

$$\oint \frac{e^{iz}}{z^2 - 2z + 2}dz = \oint \frac{e^{iz}dz}{[z-(1+i)][z-(1-i)]}$$

The root $z = 1 + i$ lies in the upper half plane, while the root $z = 1 - i$ lies in the lower half plane. We choose a contour which is a semicircle in the upper half plane, enclosing the first root. This is illustrated in Fig. 7.8.

The residue is given by

$$\begin{aligned}&\lim_{z\to i+1}\left\{[z-(1+i)]\frac{e^{iz}}{[z-(1+i)][z-(1-i)]}\right\} \\ &= \lim_{z\to i+1}\frac{e^{iz}}{z-(1-i)} \\ &= \frac{e^{-1}e^{i}}{2i}\end{aligned}$$

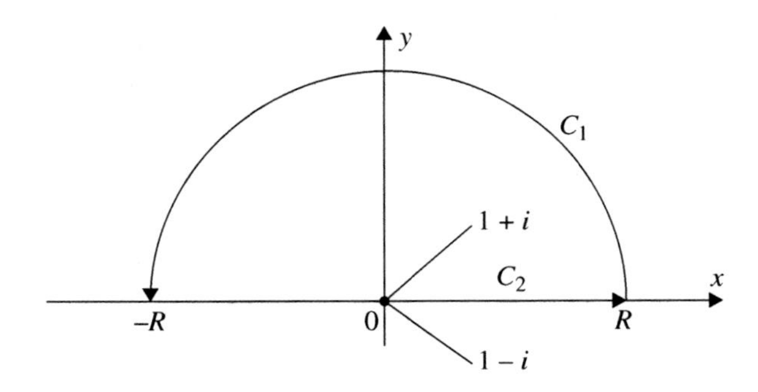

Figure 7.8 The contour used in Example 7.12.

Therefore we have

$$\oint \frac{e^{iz}dz}{z^2-2z+2} = 2\pi i\left(\frac{e^{-1}e^{i}}{2i}\right) = \frac{\pi}{e}e^{i}$$

But, using Euler's identity, we have

$$e^{i} = \cos 1 + i\sin 1$$

And so

$$\oint \frac{e^{iz}dz}{z^2-2z+2} = \frac{\pi}{e}\cos(1) + i\frac{\pi}{e}\sin(1)$$

Now, we have

$$\begin{aligned}
&\int_{-R}^{R}\frac{e^{ix}dx}{x^2-2x+2} + \int_{C_1}\frac{e^{iz}dz}{z^2-2z+2} \\
&= \int_{-R}^{R}\frac{\cos x dx}{x^2-2x+2} + i\int_{-R}^{R}\frac{\sin x dx}{x^2-2x+2} + \int_{C_1}\frac{e^{iz}dz}{z^2-2z+2} \\
&= \frac{\pi}{e}\cos(1) + i\frac{\pi}{e}\sin(1)
\end{aligned}$$

Now we let $R\to\infty$. By Jordan's lemma:

$$\int_{C_1}\frac{e^{iz}dz}{z^2-2z+2} = 0$$

So we have

$$\int_{-\infty}^{\infty} \frac{\cos x dx}{x^2 - 2x + 2} + i \int_{-\infty}^{\infty} \frac{\sin x dx}{x^2 - 2x + 2} = \frac{\pi}{e}\cos(1) + i\frac{\pi}{e}\sin(1)$$

Equating real and imaginary parts gives the result we are looking for:

$$\int_{-\infty}^{\infty} \frac{\cos x dx}{x^2 - 2x + 2} = \frac{\pi}{e}\cos(1)$$

Summary

By computing the Laurent expansion of an analytic function in a region containing one or more singularities, we were able to arrive at the residue theorem which can be used to calculate a wide variety of integrals. This includes integrals of complex functions, but the residue theorem can also be used to calculate certain classes of integrals involving functions of a real variable.

Quiz

1. Compute $\int_{\gamma} \frac{\sinh z}{z^3} dz$.
2. Compute $\int_{\gamma} \frac{\sinh z}{z^4} dz$.
3. Find the principal part of $f(z) = \frac{1}{(1+z^3)^2}$.
4. What are the singular points and residues of $\frac{\sin z}{z\left(z + \frac{5\pi}{2}\right)}$?
5. What are the singularities and residues of $\frac{\sin z}{z^2(\pi - z)}$?
6. Evaluate $\int_0^{2\pi} \frac{d\theta}{24 - 6\sin\theta}$.

7. Using the technique outlined in Example 7.9, compute $\int_0^\infty \frac{\sin^2 x}{x^2} dx$.
8. Use the residue theorem to compute $\lim_{R\to\infty} \int_{-R}^{R} \frac{dx}{1+x^2}$.
9. Compute $\int_{-\infty}^{\infty} \frac{x^2+3}{(x^2+1)(x^2+4)} dx$.
10. Compute $\int_{-\infty}^{-\infty} \frac{\cos x}{1+x^2} dx$.

CHAPTER 8

More Complex Integration and the Laplace Transform

In this chapter, we consider a few more integrals that can be evaluated using the residue theorem and then consider the Laplace transform.

Contour Integration Continued

Consider the *Fesnel integrals* which are given by

$$\int_0^\infty \cos(t^2)\,dt = \int_0^\infty \sin(t^2)\,dt = \frac{\sqrt{\pi}}{2\sqrt{2}} \tag{8.1}$$

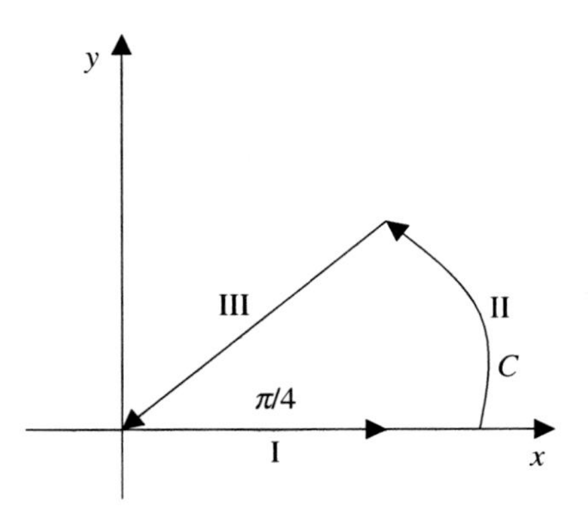

Figure 8.1 The contour C used to evaluate the integrals in Eq. (8.1).

These integrals can be evaluated by considering a wedge in the first quadrant of the complex plane with angle $\alpha = \pi/4$. This is illustrated in Fig. 8.1.

The three legs along the contour have been denoted by I, II, and III. We consider the analytic function

$$f(z) = e^{-z^2} \tag{8.2}$$

If we integrate the function in Eq. (8.2) around C, we will find it to be zero. This can be done using the residue theorem which tells us that

$$\int_C e^{-z^2}\,dz = 2\pi i \sum \text{enclosed residues}$$

This function has no singularities, a fact we can verify explicitly by writing down its series representation:

$$e^{-z^2} = 1 - z^2 + \frac{1}{2!}z^4 - \frac{1}{3!}z^6 + \cdots$$

Therefore

$$\sum \text{enclosed residues} = 0$$

And so $\int_C e^{-z^2}\,dz = 0$. Now let's try a different approach. First we break up the integral into separate integrals along each of the curves I, II, and III:

$$\int_C e^{-z^2}\,dz = \int_I e^{-z^2}\,dz + \int_{II} e^{-z^2}\,dz + \int_{III} e^{-z^2}\,dz$$

We take the radius of the wedge to be fixed (for now) at $r = R$. Now write down the polar representation:

$$z = re^{i\theta}$$

and

$$dz = dr(e^{i\theta}) + ire^{i\theta}d\theta$$

These quantities will assume different values on each leg of the contour. First consider curve I, which lies on the x axis. Along curve I, $d\theta = 0,\ \theta = 0,\ \Rightarrow z = r,\ dz = dr$.

Curve II is a circular path at radius R. So while θ varies, r is fixed. So along curve II, $r = R,\ dr = 0,\ \Rightarrow dz = iRe^{i\theta}d\theta$.

Finally, along curve III, θ is once again fixed like it was on curve I. But this time, $\theta = \pi/4,\ d\theta = 0,\ dz = e^{i\pi/4}dr$. Using these results the integral becomes

$$\int_C e^{-z^2}dz = \int_0^R e^{-r^2}dr + \int_0^{\pi/4} e^{-R^2e^{i2\theta}} iRe^{i\theta}d\theta + \int_R^0 e^{-r^2e^{i2\theta}} e^{i\pi/4}dr$$

Now, we let $R \to \infty$. On curve II, as $R \to \infty$ we have

$$\lim_{R\to\infty}\int_0^{\pi/4} e^{-R^2e^{i2\theta}} iRe^{i\theta}d\theta = 0$$

because $e^{-R^2e^{i2\theta}} \to 0$ is much faster than $R \to \infty$. Earlier, we found out that $\int_C e^{-z^2}dz = 0$. Then we wrote $\int_C e^{-z^2}dz = \int_0^R e^{-r^2}dr + \int_0^{\pi/4} e^{-R^2e^{i2\theta}} iRe^{i\theta}d\theta + \int_R^0 e^{-r^2e^{i2\theta}} e^{i\pi/4}dr$ but just noted that as $R \to \infty \int_0^{\pi/4} e^{-R^2e^{i2\theta}} iRe^{i\theta}d\theta = 0$. So we're left with

$$\lim_{R\to\infty}\left\{\int_0^R e^{-r^2}dr - \int_0^R e^{-r^2e^{i2\theta}} e^{i\pi/4}dr\right\} = 0$$

The first integral can be looked up in a table:

$$\int_0^\infty e^{-r^2}dr = \frac{\sqrt{\pi}}{2} \tag{8.3}$$

Using Euler's identity:

$$e^{i\pi/4} = \cos\pi/4 + i\sin\pi/4 = \frac{1+i}{\sqrt{2}}$$

Therefore

$$\frac{\sqrt{\pi}}{2} = \frac{1+i}{\sqrt{2}}\int_0^\infty e^{-ir^2}\,dr = \frac{1+i}{\sqrt{2}}\int_0^\infty \{\cos(r^2) + i\sin(r^2)\}\,dr$$
$$= \frac{1}{\sqrt{2}}\int_0^\infty \cos(r^2)\,dr + \frac{1}{\sqrt{2}}\int_0^\infty \sin(r^2)\,dr + \frac{i}{\sqrt{2}}\int_0^\infty \{\cos(r^2) - \sin(r^2)\}\,dr$$

Now equate real and imaginary parts. Clearly $\text{Im}(\sqrt{\pi}/2) = 0$, so it follows that

$$\frac{i}{\sqrt{2}}\int_0^\infty \{\cos(r^2) - \sin(r^2)\}\,dr = 0$$
$$\Rightarrow \int_0^\infty \cos(r^2)\,dr = \int_0^\infty \sin(r^2)\,dr$$

Equating real parts and using $\int_0^\infty \cos(r^2)\,dr = \int_0^\infty \sin(r^2)\,dr$ gives

$$\frac{\sqrt{\pi}}{2} = \frac{1}{\sqrt{2}}\int_0^\infty \cos(r^2)\,dr + \frac{1}{\sqrt{2}}\int_0^\infty \sin(r^2)\,dr = \frac{2}{\sqrt{2}}\int_0^\infty \cos(r^2)\,dr$$
$$\Rightarrow \int_0^\infty \cos(r^2)\,dr = \frac{\sqrt{\pi}}{2\sqrt{2}}$$

Next, we consider the integral $\int_0^x [dz/(1+z^2)]$, which you've already seen in Chap. 6. Here, we show how to do it using natural logarithms and a little trick. First we factor the integrand:

$$\int_0^x \frac{dz}{1+z^2} = \int_0^x \frac{dz}{(z-i)(z+i)} = \frac{1}{2i}\int_0^x \frac{dz}{z-i} - \frac{1}{2i}\int_0^x \frac{dz}{z+i}$$

The last step used partial fraction decomposition. These integrals can be solved readily to give

$$\int_0^x \frac{dz}{1+z^2} = \frac{1}{2i}\{\ln(z-i) - \ln(z+i)\}\Big|_0^x$$

In Fig. 8.2, we draw both lines on a triangle which will help us evaluate the integral.

Notice from Fig. 8.2 that

$$\tan\theta = \frac{z}{|i|} = z \qquad \Rightarrow \theta = \tan^{-1} z$$

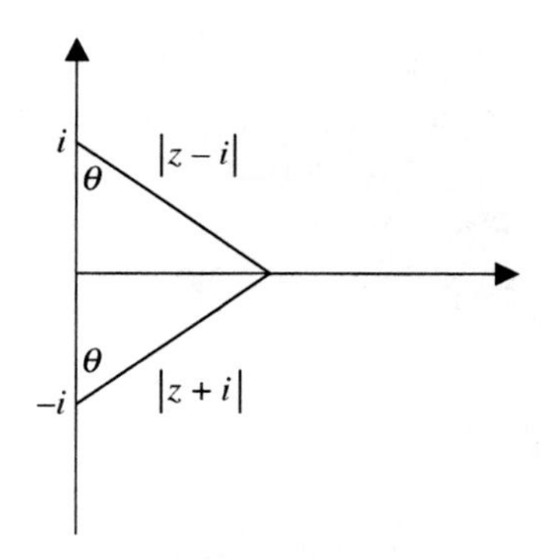

Figure 8.2 We draw triangles to evaluate $\int_0^x [dz/(1+z^2)] = (1/2i)\{\ln(z-i) - \ln(z+i)\}$ at the upper and lower limits.

Moreover we have

$$\begin{aligned}\ln(z-i) - \ln(z+i) &= \ln|z-i| + i\arg(z-i) - \{\ln|z+i| + i\arg(z+i)\} \\ &= i\arg(z-i) - i\arg(z+i) \\ &= 2i\theta = 2i\tan^{-1} z\end{aligned}$$

Therefore the integral can be evaluated in the following way:

$$\int_0^x \frac{dz}{1+z^2} = \frac{1}{2i}\{2i\tan^{-1} z\}\Big|_0^x = \tan^{-1} x$$

The Laplace Transform

So far we've seen how complex integration can make many integrals that seem impossible to evaluate much easier to tackle. Now we turn our attention to the notion of a *transform*, which is a method that takes the representation of a function in terms of one variable (say time or position) and represents it in terms of a different variable like frequency. This type of mathematical operation leads to simplification of many tasks like solving differential equations, which can be turned into algebraic relationships. The first transform we will investigate is the *Laplace transform*.

The Laplace transform is a useful mathematical tool that converts functions of time into functions of a complex variable denoted by s. This technique is very useful because the Laplace transform allows us to convert ordinary differential equations into algebraic equations which are usually easier to solve.

A Laplace transform can be used to transform a function of time t or position x into a function of the complex variable s. In the definitions that follow, we will stick to considering functions of time, but keep in mind that x could equally well be used in place of t. We write the complex variable s in terms of real and imaginary parts as follows:

$$s = \sigma + i\omega \tag{8.4}$$

where $\sigma = \text{Re}(s)$ and $\omega = \text{Im}(s)$ are real variables. The definition of the Laplace transform is given in terms of an integral. The Laplace transform $F(s)$ of a function $f(t)$ is given by

$$F(s) = \int_{-\infty}^{\infty} f(t) e^{-st}\, dt \tag{8.5}$$

This can be written in an abstract form as

$$F(s) = \mathfrak{L}\{f(t)\} \tag{8.6}$$

where $\mathfrak{L}\{\bullet\}$ is the Laplace transform viewed as an operator acting on the function $f(t)$. Let's compute a few Laplace transforms using Eq. (8.5).

EXAMPLE 8.1

Find the Laplace transform of $f(t) = \theta(t)$, where $\theta(t) = \begin{cases} 1 t > 0 \\ 0 t \leq 0 \end{cases}$ (see Fig. 8.3).

SOLUTION

Inserting this function into the defining integral in Eq. (8.5) we find

$$F(s) = \int_{-\infty}^{\infty} \theta(t) e^{-st}\, dt = \int_{0}^{\infty} e^{-st}\, dt = -\frac{1}{s} e^{-st} \Big|_0^{\infty} = \frac{1}{s}$$

Therefore we have the *Laplace transform pair*

$$\mathfrak{L}\{\theta(t)\} = \frac{1}{s}$$

EXAMPLE 8.2

Find the Laplace transform of (see Fig. 8.4)

$$f(t) = e^{-at} u(t)$$

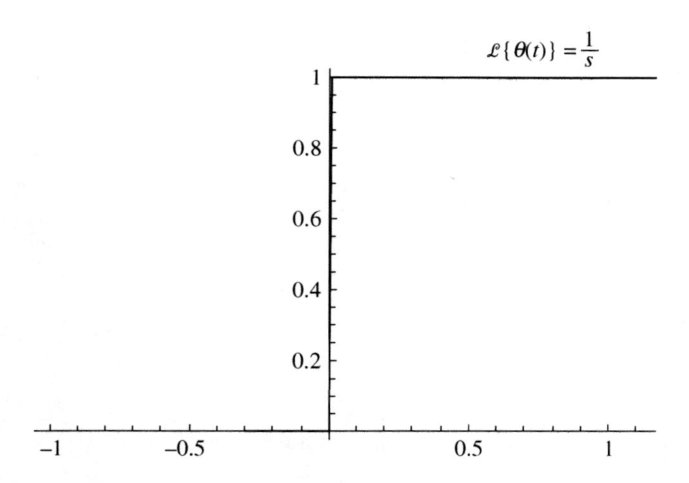

Figure 8.3 In Example 8.1, we find the Laplace transform of $f(t) = \theta(t)$. This function is called the *Heaviside* or *unit step function*.

SOLUTION

Again using the defining integral we have

$$F(s) = \int_{-\infty}^{\infty} e^{-at}\theta(t)e^{-st}\,dt = \int_{0}^{\infty} e^{-at}e^{-st}\,dt = \int_{0}^{\infty} e^{-(a+s)t}\,dt$$

$$= -\frac{1}{(s+a)}e^{-(s+a)t}\Big|_0^{\infty} = \frac{1}{(s+a)}$$

So we have the Laplace transform pair

$$\mathcal{L}\{e^{-at}\theta(t)\} = \frac{1}{s+a}$$

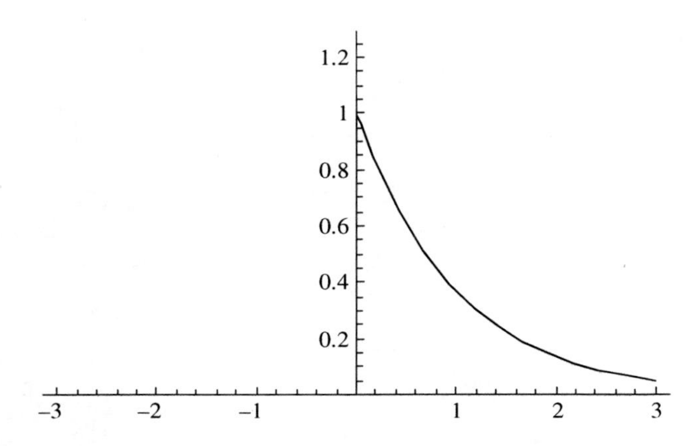

Figure 8.4 In Example 8.2, we compute the Laplace transform of $f(t) = e^{-at}u(t)$.

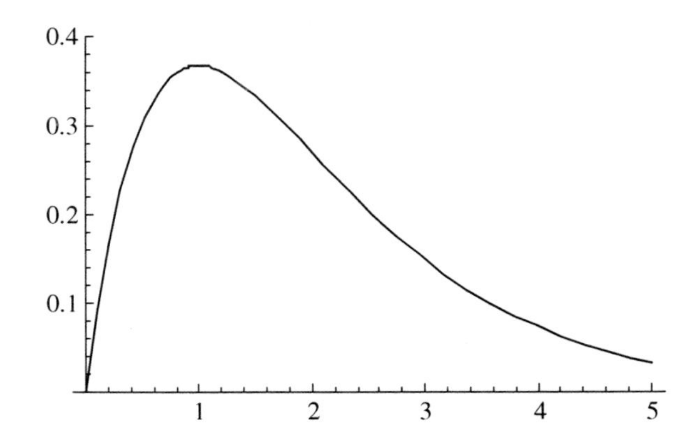

Figure 8.5 In Example 8.3 we find the Laplace transform of $x(t) = te^{-at}\theta(t)$.

EXAMPLE 8.3

Find the Laplace transform of $x(t) = te^{-at}\theta(t)$.

SOLUTION

The function is displayed in Fig. 8.5. Proceeding, we have

$$F(s) = \int_{-\infty}^{\infty} te^{-at}\theta(t)e^{-st}\,dt = \int_{0}^{\infty} te^{-at}e^{-st}\,dt = \int_{0}^{\infty} te^{-(a+s)t}\,dt$$

This is a familiar integral to most readers. No doubt many recall that this integral can be done using integration by parts. It's always good to review so let's quickly go through the process to refresh our memories. We start by recalling the integration by parts formula:

$$\int u\,dv = uv - \int v\,du$$

In the case at hand, we set

$$u = t \qquad \Rightarrow du = dt$$

$$dv = e^{-(s+a)t}dt \qquad \Rightarrow v = -\frac{1}{s+a}e^{-(s+a)t}$$

Therefore, applying the integration by parts formula we obtain

$$\begin{aligned} F(s) &= -\frac{1}{s+a}te^{-(s+a)t}\Big|_0^{\infty} + \frac{1}{s+a}\int_0^{\infty} e^{-(s+a)t}dt \\ &= \frac{1}{s+a}\int_0^{\infty} e^{-(s+a)t}dt \\ &= -\frac{1}{(s+a)^2}e^{-(s+a)t}\Big|_0^{\infty} = \frac{1}{(s+a)^2} \end{aligned}$$

So, we have derived the Laplace transform pair:

$$\mathcal{L}\{te^{-at}\theta(t)\} = \frac{1}{(s+a)^2}$$

EXAMPLE 8.4

Let $\alpha > -1$. The factorial function is defined by $\alpha! = \int_0^\infty e^{-t}t^\alpha dt$. Find $(-1/2)!$ and then show that the Laplace transform of t^β is given by $\beta!/s^{\beta+1}$.

SOLUTION

Using the definition $\alpha! = \int_0^\infty e^{-t}t^\alpha dt$ we have

$$(-1/2)! = \int_0^\infty e^{-t}t^{-1/2}dt$$

Let $\sqrt{t} = x$. Then $[1/(2\sqrt{t})]dt = x, t = x^2$ and

$$(-1/2)! = \int_0^\infty e^{-t}t^{-1/2}dt$$
$$= 2\int_0^\infty e^{-x^2}dx = 2\left(\frac{1}{2}\right)\int_{-\infty}^\infty e^{-x^2}dx = \sqrt{\pi}$$

Using Eq. (8.5) the Laplace transform of t^β is

$$\int_0^\infty e^{-st}t^\beta dt$$

Let $r = st$. Then $dr = sdt,\ t = r/s \Rightarrow$

$$\int_0^\infty e^{-st}t^\beta dt = \int_0^\infty e^{-r}\left(\frac{r}{s}\right)^\beta \frac{dr}{s} = \frac{1}{s^{\beta+1}}\int_0^\infty e^{-r}r^\beta dr = \frac{\beta!}{s^{\beta+1}}$$

IMPORTANT PROPERTIES OF THE LAPLACE TRANSFORM

The Laplace transform is a linear operation. This follows readily from the defining integral. Suppose that $F_1(s) = \mathcal{L}\{f_1(t)\}$ and $F_2(s) = \mathcal{L}\{f_2(t)\}$. Also, let α, β be two constants. Then

$$\mathcal{L}\{\alpha f_1(t) + \beta f_2(t)\} = \int_{-\infty}^\infty [\alpha f_1(t) + \beta f_2(t)]\ e^{-st}dt \qquad (8.7)$$

Now, we can just use the linearity properties of the integral. We have

$$\begin{aligned}\int_{-\infty}^{\infty}[\alpha f_1(t)+\beta f_2(t)]e^{-st}dt &= \int_{-\infty}^{\infty}[\alpha f_1(t)e^{-st}+\beta f_2(t)e^{-st}]dt\\ &= \int_{-\infty}^{\infty}\alpha f_1(t)e^{-st}\,dt+\int_{-\infty}^{\infty}\beta f_2(t)e^{-st}\,dt\\ &= \alpha\int_{-\infty}^{\infty}f_1(t)e^{-st}\,dt+\beta\int_{-\infty}^{\infty}f_2(t)e^{-st}\,dt\\ &= \alpha F_1(s)+\beta F_2(s)\end{aligned}$$

The next property we want to look at is *time scaling*. Suppose that we have a continuous function of time $f(t)$ and some constant $a>0$. Given that

$$F(s)=\int_{-\infty}^{\infty}f(t)e^{-st}\,dt$$

what is the Laplace transform of the time scaled function $f(at)$? The defining integral is

$$\mathcal{L}\{f(at)\}=\int_{-\infty}^{\infty}f(at)e^{-st}dt$$

Let's fix this up with a simple change of variables. Let

$$u=at \qquad \Rightarrow du=adt$$

Furthermore, we can write

$$t=\frac{u}{a}$$

So we have

$$\mathcal{L}\{f(at)\}=\int_{-\infty}^{\infty}f(at)e^{-st}dt=\frac{1}{a}\int_{-\infty}^{\infty}f(u)e^{-su/a}du$$

Let's write the argument of the exponential in a more suggestive way

$$-\frac{su}{a}=-\left(\frac{s}{a}\right)u=-\theta u$$

where for the moment we have defined another new variable $\theta=s/a$. This change makes the above integral look just like a plain old Laplace transform. Since the

integration variable u is just a "dummy" variable, we can call it anything we like. Let's put in the aforementioned changes and also let $u \to t$

$$\mathcal{L}\{f(at)\} = \frac{1}{a}\int_{-\infty}^{\infty} f(u)e^{-su/a}du = \frac{1}{a}\int_{-\infty}^{\infty} f(u)e^{-\theta u}du$$
$$= \frac{1}{a}\int_{-\infty}^{\infty} f(t)e^{-\theta t}du = \frac{1}{a}F(\theta)$$

Now we change back $\theta = s/a$ and we have discovered that

$$\mathcal{L}\{f(at)\} = \frac{1}{a}F\left(\frac{s}{a}\right)$$

More generally, if we let a assume negative values as well, this relation is written as

$$\mathcal{L}\{f(at)\} = \frac{1}{|a|}F\left(\frac{s}{a}\right) \tag{8.8}$$

The next property we wish to consider is *time shifting*. Suppose that

$$F(s) = \mathcal{L}\{f(t)\}$$

What is the Laplace transform of $x(t-t_o)$? Using the definition of the Laplace transform we have

$$\mathcal{L}\{f(t-t_o)\} = \int_{-\infty}^{\infty} f(t-t_o)e^{-st}\,dt$$

Once again, we can proceed with a simple change of variables. We let $u = t - t_o$ from which it follows immediately that $du = dt$. Then

$$\mathcal{L}\{f(t-t_o)\} = \int_{-\infty}^{\infty} f(u)e^{-s(u+t_o)}\,du$$
$$= \int_{-\infty}^{\infty} f(u)e^{-su}e^{-st_o}\,du$$

Notice that s and t_o are not integration variables, so given any term that is a function of these variables alone, we can just pull it outside the integral. This gives

$$\mathcal{L}\{f(t-t_o)\} = e^{-st_o}\int_{-\infty}^{\infty} f(u)e^{-su}\,du = e^{-st_o}\int_{-\infty}^{\infty} f(t)e^{-st}\,dt = e^{-st_o}F(s)$$

We conclude that the effect of a time shift by t_o is to multiply the Laplace transform by e^{-st_o}.

DIFFERENTIATION

When we consider the derivative of a function of time, we encounter one of the most useful properties of the Laplace transform which makes it well suited to use when solving ordinary differential equations. Starting with the defining integral consider the Laplace transform

$$\mathcal{L}\left\{\frac{df}{dt}\right\}$$

We only consider functions that vanish when $t < 0$. This is just

$$\int_0^\infty \frac{df}{dt} e^{-st}\, dt$$

Let's use our old friend integration by parts to move the derivative away from $f(t)$. We obtain

$$\int_0^\infty \frac{df}{dt} e^{-st}\, dt = f(t)e^{-st}\Big|_0^\infty + s\int_0^\infty f(t)e^{-st}\, dt$$

We consider the boundary term first. Clearly $f(t)e^{-st}$ goes to zero at the upper limit because the decaying exponential goes to zero as $t \to \infty$. Therefore

$$f(t)e^{-st}\Big|_0^\infty = -f(0)$$

Now take a look at the integral in the second term. This is nothing other than the Laplace transform of $f(t)$. So, we find that

$$\mathcal{L}\left\{\frac{df}{dt}\right\} = -f(0) + sF(s) \tag{8.9}$$

Now let's consider differentiation with respect to s. That is:

$$\frac{d}{ds}[F(s)] = \frac{d}{ds}\int_{-\infty}^\infty f(t)\, e^{-st}\, dt$$

The integration is with respect to t, so it seems fair enough that we can slide the derivative with respect to s on inside the integral. This gives

$$\frac{d}{ds}[F(s)] = \int_{-\infty}^\infty \frac{d}{ds}[f(t)\, e^{-st}]\, dt = \int_{-\infty}^\infty f(t)\frac{d}{ds}(e^{-st})dt = -\int_{-\infty}^\infty t\, f(t)e^{-st}\, dt$$

We can move the minus sign to the other side, that is:

$$-\frac{dF}{ds} = \int_{-\infty}^{\infty} t\,f(t)e^{-st}\,dt$$

This tells us that the Laplace transform of $t\,f(t)$ is given by

$$\mathcal{L}\{t\,f(t)\} = -\frac{dF}{ds} \tag{8.10}$$

EXAMPLE 8.5
Given that

$$\mathcal{L}\{\cos \beta t u(t)\} = \frac{s}{s^2 + \beta^2}$$

Find $\mathcal{L}\{t \cos \beta t u(t)\}$ (see Fig. 8.6).

SOLUTION
We obtain the result by computing the derivative of $s/(s^2 + \beta^2)$ and adding a minus sign. First we recall that the derivative of a quotient is given by

$$\left(\frac{f}{g}\right)' = \frac{f'g - g'f}{g^2}$$

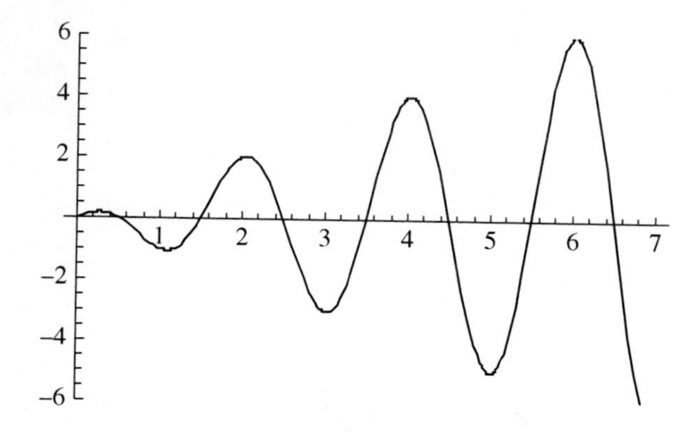

Figure 8.6 In Example 8.5, we compute the Laplace transform of $t \cos \beta t u(t)$. The function is shown here with $\beta = \pi$.

In the case of $s/(s^2+\beta^2)$ we have

$$f = s \qquad \Rightarrow f' = 1$$
$$g = s^2+\beta^2 \qquad \Rightarrow g' = 2s$$

And so

$$\left(\frac{f}{g}\right)' = \frac{f'g - g'f}{g^2} = \frac{(s^2+\beta^2) - 2s(s)}{(s^2+\beta^2)^2} = \frac{\beta^2 - s^2}{(s^2+\beta^2)^2}$$

Applying Eq. (8.10) we add a minus sign and find that.

EXAMPLE 8.6

Find the solution of

$$\frac{dy}{dt} = A\cos t$$

for $t \geq 0$ where A is a constant and $y(0)=1$. See Fig. 8.7.

SOLUTION

This is a very simple ordinary differential equation (ODE) and it can be verified by integration that the solution is $y(t) = 1 + A\sin t$ (see Fig. 8.6). Since this is an easy ODE to solve it's a good one to use to illustrate the method of the Laplace transform. Taking the Laplace transform of the left side, we have

$$\mathcal{L}\left\{\frac{dy}{dt}\right\} = -y(0) + sY(s) = -1 + sY(s)$$

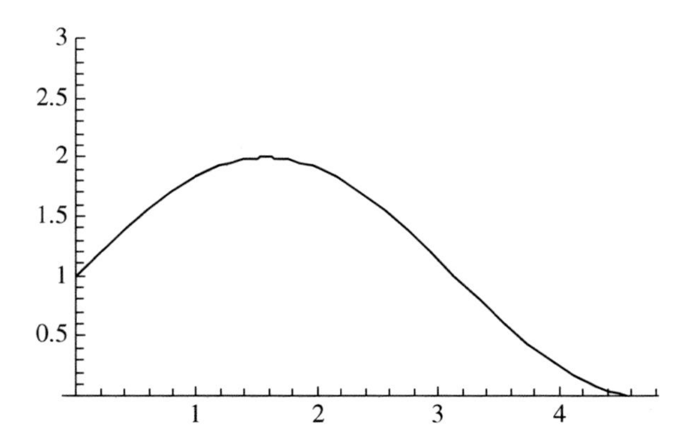

Figure 8.7 A plot of the solution to $dy/dt = A\cos t$ with $A = 1$ and $y(0) = 1$.

In the chapter quiz, you will show that

$$\mathcal{L}\{\cos \beta t u(t)\} = \frac{s}{s^2 + \beta^2}$$

This tells us that the Laplace transform of the right-hand side of the differential equation is

$$\mathcal{L}\{A \cos t\} = A \frac{s}{s^2 + 1}$$

(remember, $t \geq 0$ was specified in the problem, so we don't need to explicitly include the unit step function). Equating both sides gives us an equation we can solve algebraically

$$-1 + sY(s) = A \frac{s}{s^2 + 1}$$

Adding 1 to both sides we obtain

$$sY(s) = 1 + A \frac{s}{s^2 + 1}$$

Now we divide through by s, giving an expression for the Laplace transform of $y(t)$

$$Y(s) = \frac{1}{s} + A \frac{1}{s^2 + 1}$$

Earlier we found that

$$\mathcal{L}\{u(t)\} = \frac{1}{s}$$

Since it has been specified that $t \geq 0$, this is the same as stating that

$$\mathcal{L}\{1\} = \frac{1}{s}$$

In the chapter quiz, you will show that

$$F(s) = \frac{\beta}{s^2 + \beta^2}$$

is the Laplace transform of $f(t) = \sin(\beta t)\, u(t)$. Putting these results together, by inspection of

$$Y(s) = \frac{1}{s} + A\frac{1}{s^2+1}$$

We conclude that

$$y(t) = 1 + A\sin t$$

EXAMPLE 8.7

Abel's integral equation is

$$f(x) = \int_0^x \frac{\phi(t)}{\sqrt{x-t}}\, dt$$

The function $f(x)$ is given while ϕ is an unknown. Using the Laplace transform, show that

$$\sqrt{\frac{\pi}{s}}\, F(s) = \frac{\pi}{s}\Phi(s)$$

And that we can solve for the unknown using $\phi(x) = (1/\pi)\int_0^x [f(t)/\sqrt{x-t}]\, dt$.

SOLUTION

The integral of a function of time multiplied by another function of time which is shifted is called *convolution*. This is defined as

$$\int f(t)\, g(x-t)\, dt = f * g \tag{8.11}$$

It can be shown that the Laplace transform turns convolution into multiplication:

$$\mathfrak{L}\{f * g\} = F(s)G(s) \tag{8.12}$$

If we take

$$g(x) = \frac{1}{\sqrt{x}}$$

Then

$$f(x) = \int_0^x \frac{\phi(t)}{\sqrt{x-t}}\,dt = \phi * g$$

Using the results of Example 8.3, we have

$$\mathcal{L}\{\phi * g\} = \Phi(s)\frac{(-1/2)!}{\sqrt{s}} = \Phi(s)\sqrt{\frac{\pi}{s}}$$

It follows that $\sqrt{\pi/s}\,F(s) = (\pi/s)\ \Phi(s)$. That is, $\Phi(s) = \sqrt{s/\pi}\,F(s)$. Now, the Laplace transform has an inverse. Since we know that the Laplace transform of a convolution is a product and we know what the Laplace transform of a derivative is, then

$$\begin{aligned}
\mathcal{L}^{-1}\{\Phi(s)\} &= \mathcal{L}^{-1}\left\{\sqrt{\frac{s}{\pi}}F(s)\right\} \\
&= \mathcal{L}^{-1}\left\{\frac{1}{\pi}s\sqrt{\frac{\pi}{s}}F(s)\right\} \\
&= \frac{1}{\pi}\mathcal{L}^{-1}\left\{s\sqrt{\frac{\pi}{s}}F(s)\right\} \\
&= \frac{1}{\pi}\frac{d}{dx}\left[f(x) * \frac{1}{\sqrt{x}}\right] = \frac{1}{\pi}\frac{d}{dx}\int_0^x \frac{f(t)}{\sqrt{x-t}}\,dt
\end{aligned}$$

The Bromvich Inversion Integral

The inverse Laplace transform is defined as follows. Consider the function $g(t) = e^{-ct}f(t)$ where c is a real constant. Now, using Fourier transforms:

$$\begin{aligned}
g(t) &= \int_{-\infty}^{\infty} e^{i\omega t}G(\omega)\,d\omega \\
&= \int_{-\infty}^{\infty} e^{i\omega t}\left(\frac{1}{2\pi}\int_{-\infty}^{\infty} e^{-i\omega\tau}g(\tau)\,d\tau\right)d\omega \\
&= \int_{-\infty}^{\infty} e^{i\omega t}\int_{0}^{\infty} e^{-i\omega\tau}e^{-c\tau}f(\tau)\,d\tau
\end{aligned}$$

Using $g(t) = e^{-ct} f(t)$, we obtain the relation

$$f(t) = \frac{1}{2\pi} \int_{-\infty}^{\infty} e^{(c+i\omega)t} \int_{0}^{\infty} e^{-(c+i\omega)\tau} f(\tau)\, d\tau\, d\omega$$

Now let $z = c + i\omega$, then

$$f(t) = \frac{1}{2\pi i} \int_{c-i\infty}^{c+i\infty} e^{zt} \int_{0}^{\infty} e^{-z\tau} f(\tau)\, d\tau\, dz$$

And we obtain the Bromwich inversion integral:

$$f(t) = \frac{1}{2\pi i} \int_{c-i\infty}^{c+i\infty} e^{zt} F(z)\, dz \tag{8.13}$$

The *Bromwich contour* is a line running up and down the y axis from $c - iR$ to $c + iR$ (then we let $R \to \infty$).

EXAMPLE 8.8

Find the inverse Laplace transform of $F(s) = 1/(s^2 + \omega^2)$.

SOLUTION

Notice that

$$F(s) = \frac{1}{s^2 + \omega^2} = \frac{1}{(s - i\omega)(s + i\omega)}$$

So this function has two singularities (simple poles) at $s = \pm i\omega$. The inversion integral in Eq. (8.13) in this case becomes

$$f(t) = \frac{1}{2\pi i} \int_{c-i\infty}^{c+i\infty} \frac{e^{st}}{(s - i\omega)(s + i\omega)} ds$$

The residue at $s = +i\omega$ is

$$(s - i\omega) \frac{e^{st}}{(s - i\omega)(s + i\omega)} \bigg|_{s = i\omega} = \frac{e^{i\omega t}}{2i\omega}$$

The other residue is

$$(s+i\omega)\frac{e^{st}}{(s-i\omega)(s+i\omega)}\bigg|_{s=i\omega}=-\frac{e^{-i\omega t}}{2i\omega}$$

The integral is already divided by $2\pi i$, so by the residue theorem

$$f(t)=\sum res=\frac{e^{i\omega t}}{2i\omega}-\frac{e^{-i\omega t}}{2i\omega}=\frac{\sin\omega t}{\omega}$$

Summary

In this chapter, we explored more complex integrals and introduced the Laplace transform, a tool which can be used to solve differential equations algebraically. The inverse Laplace transform is defined using a contour integral called the Bromvich inversion integral.

Quiz

1. Calculate the Laplace transform of $\cos\omega t$.
2. Find the Laplace transform of $\cosh at$ where a is a constant.
3. Using the Bromvich inversion integral, find the inverse Laplace transform of $\frac{e^{-k\sqrt{s}}}{\sqrt{s}}$.
4. Using the Bromvich inversion integral, find the inverse Laplace transform of $\frac{s}{s^2+\omega^2}$.
5. Using the Bromvich inversion integral and $\alpha! = \int_0^\infty e^{-t}t^\alpha dt$, find the inverse Laplace transform of $F(s)=s^{-\alpha}$.

CHAPTER 9

Mapping and Transformations

In this chapter, we will introduce a few of the techniques that can be used to transform a region of the complex plane into another different region of the complex plane. You may want to do this because it will be convenient for a given problem you're solving. There are many types of transformations that can be applied in the limited space we have, we won't be able to cover but a small fraction of them. Our purpose here is to introduce you to a few of the common transformations used and get you used to the concepts involved.

Let us define two complex planes. The first is the z plane defined by the coordinates x and y. We will now introduce a second plane, which we call the w plane, defined by two coordinates that are denoted by u and v. *Mapping* is a transformation between points in the z plane and points in the w plane. This is illustrated in Fig. 9.1.

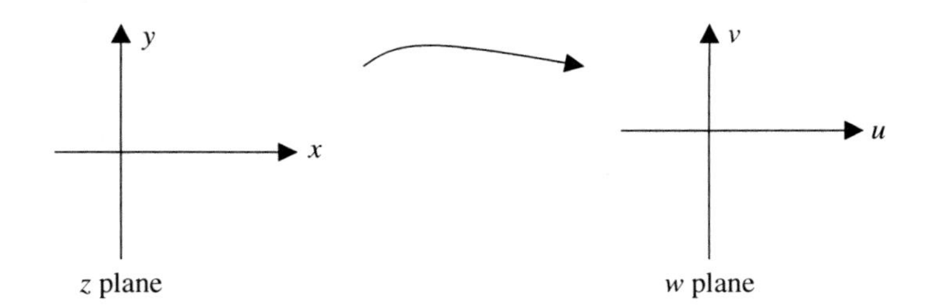

Figure 9.1 Mapping is a transformation of points in the z plane to points in a new w plane.

Linear Transformations

A *linear transformation* is one that relates w to z by a linear equation of the form

$$w = \alpha z + \beta \tag{9.1}$$

where α and β are complex constants. Consider for a moment the transformation

$$w = \alpha z$$

To see the effect of this transformation, we can write each factor in the polar representation. Let

$$\alpha = ae^{i\phi}$$

and as usual, denote $z = re^{i\theta}$. Then

$$w = \alpha z = (ae^{i\phi})(re^{i\theta}) = are^{i(\phi+\theta)}$$

If $a > 1$, then the transformation expands the radius vector of z through the transformation $r \to ar$. If $a < 1$, then the transformation contracts the radius vector of z as $r \to ar = (1/b)\,r$, where $b > 1$. The transformation rotates the point z by an angle given by

$$\phi = \arg(\alpha)$$

about the origin. The w plane is defined by the coordinate $w = u + iv$.

EXAMPLE 9.1

Explain what the transformation $w = iz$ does to the line $y = x + 2$ in the x-y plane.

SOLUTION

Note that

$$w = iz = i(x + iy) = -y + ix$$

So we have the relations

$$u = -y$$
$$v = x$$

Hence

$$y = x + 2 \Rightarrow -u = v + 2$$

That is, the line is transformed to

$$v = -u - 2$$

This linear transformation maps one line into another one, as illustrated in Fig. 9.2.

EXAMPLE 9.2

Consider the transformation $w = (1 + i)z$ on the rectangular region shown in Fig. 9.3.

SOLUTION

Notice that

$$1 + i = \sqrt{2}\left(\frac{1 + i}{\sqrt{2}}\right) = \sqrt{2}(\cos\pi/4 + i\sin\pi/4) = \sqrt{2}e^{i\pi/4}$$

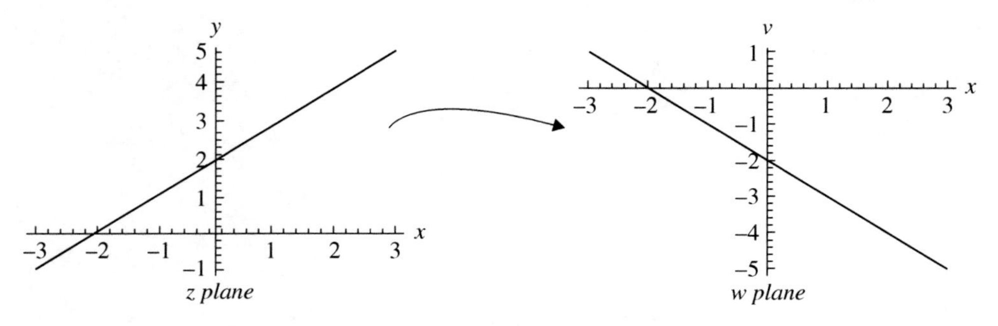

Figure 9.2 The transformation $w = iz$ described in Example 9.1.

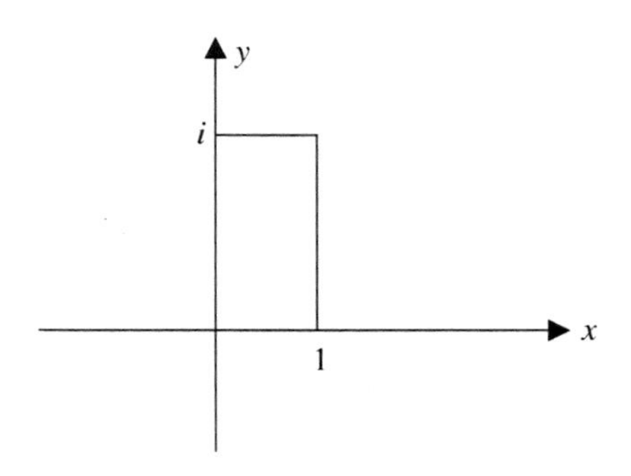

Figure 9.3 A rectangular region to be transformed by $w = (1+i)z$ in Example 9.2.

This tells us that the transformation will stretch lengths by $\sqrt{2}$ and rotate points in a counterclockwise direction about the origin by the angle $\pi/4$. The transformed points are

$$w = (1+i)z = (1+i)(x+iy) = x - y + i(x+y)$$
$$\Rightarrow u = x - y \qquad v = x + y$$

So the points on the rectangle are transformed according to

$$(1,0) \to (1,i)$$
$$(1,i) \to (0,2i)$$
$$(0,i) \to (-1,i)$$
$$(0,0) \to (0,0)$$

The transformed rectangle is illustrated in Fig. 9.4.

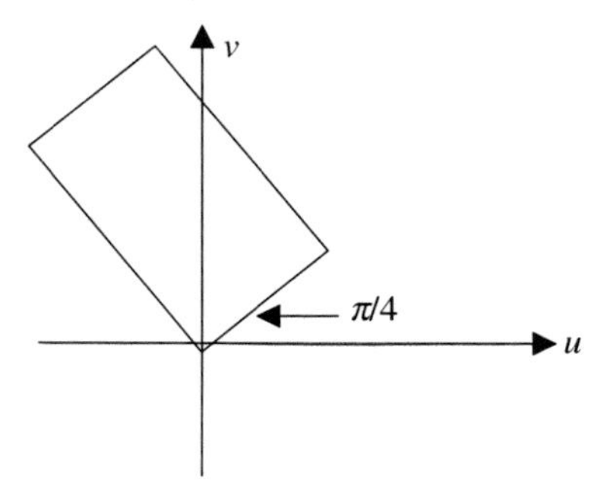

Figure 9.4 The rectangle in Fig. 9.3 transformed by $w = (1+i)z$. It is rotated by $\pi/4$ about the origin and lengths are increased by $\ell \to \sqrt{2}\ell$.

If the transformation is of the form $w = \alpha z + \beta$, the effect is to translate the region to the left or to the right by the magnitude of the real part of β, and up or down by the magnitude of the imaginary part of β. Consider the transformation

$$w = (1+i)z + 3$$

on the rectangular region shown in Fig. 9.3. The effect of this transformation is to first rotate and expand lengths, and then to translate along the real axis. The points at the four corners of the rectangle are transformed according to

$$(1,0) \to (4,i)$$
$$(1,i) \to (3,2i)$$
$$(0,i) \to (2,i)$$
$$(0,0) \to (3,0)$$

The transformed rectangular region looks something like that in Fig. 9.5.

Now let's consider a square region in the z plane defined by $0 \leq x \leq 1, 0 \leq y \leq 1$. Consider the transformation

$$w = z + 1 + i$$

All this does is shift the square over to the right one unit and up one unit. This is illustrated in Fig. 9.6.

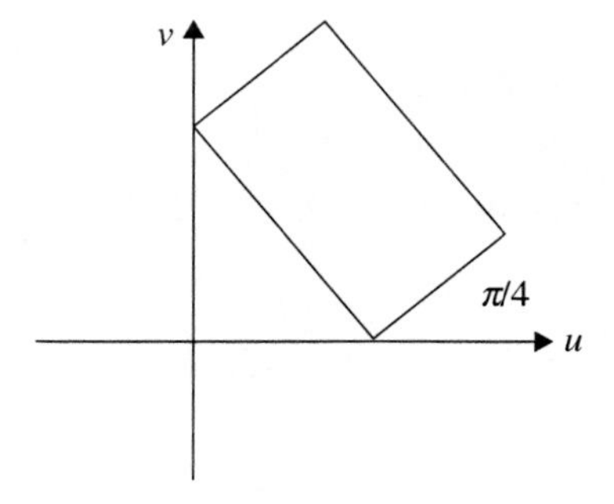

Figure 9.5 Adding a translation along the real axis to the transformation of Example 9.2.

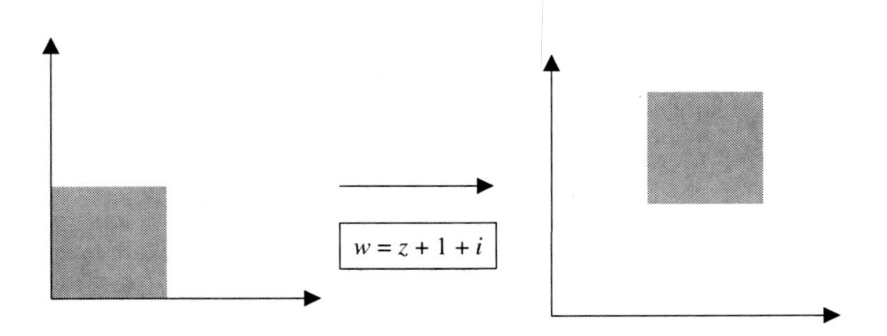

Figure 9.6 Consider a region that is a square by the origin. The transformation $w = z + 1 + i$ shifts the square up and over.

The Transformation z^n

The transformation $w = z^n$ changes lengths according to $r \to r^n$ and increases angles by a factor of n. Letting $z = re^{i\theta}$, we see that the transformation is given in polar coordinates as

$$w = z^n = (re^{i\theta})^n = r^n e^{in\theta}$$

Before considering a mapping of a region, we let $x = a$ be a vertical line in the plane. The transformation $w = z^2$ gives

$$w = z^2 = (a + iy)^2 = a^2 - y^2 + i2ay$$
$$\Rightarrow u = a^2 - y^2 \qquad v = 2ay$$

This allows us to set

$$y = \frac{v}{2a}$$

Hence the transformation is a parabola in the w plane described by the equation

$$u = a^2 - \left(\frac{v}{2a}\right)^2$$

EXAMPLE 9.3

Consider the transformation of the quarter plane as shown in Fig. 9.7 under the mapping $w = z^2$.

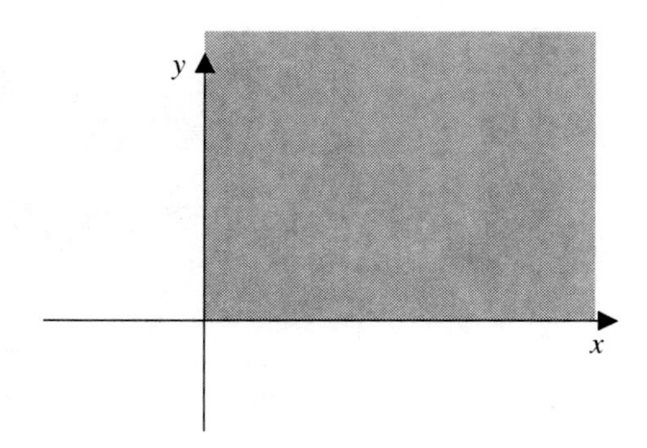

Figure 9.7 In Example 9.3 we apply the transformation $w = z^2$ to the quarter plane defined by $0 \le x, 0 \le y$.

SOLUTION

We have seen that the effect of $w = z^n$ is to increase angles by a factor of n. Indeed, in this case

$$w = z^2 = (re^{i\theta})^2 = r^2 e^{i2\theta}$$

Hence, angles are doubled. So the angle $\pi/2$ that defines the quarter plane is expanded as $\pi/2 \to \pi$. That is, the quarter plane is mapped to the upper half plane by this transformation. This is shown in Fig. 9.8.

Generally speaking, consider a triangular region in the x-y plane with angle $\theta = \pi/n$. The mapping $w = z^n$ maps this region to the half plane. This is illustrated in Fig. 9.9.

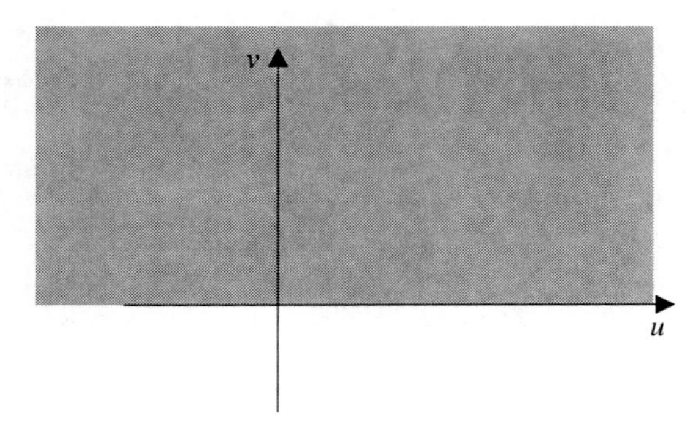

Figure 9.8 The transformation $w = z^n$ has mapped the quarter plane to the half plane.

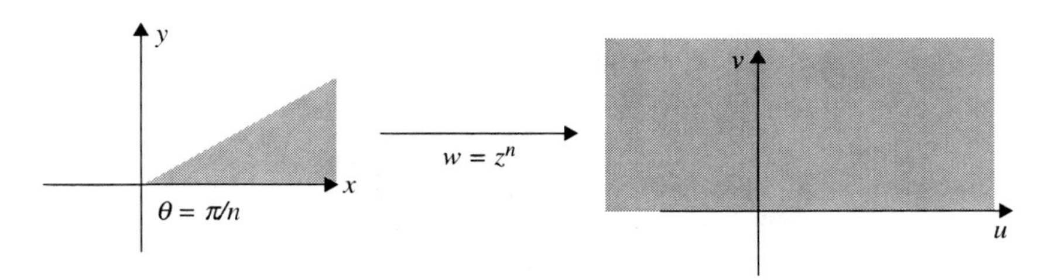

Figure 9.9 A infinite sector defined by a triangular wedge with $\theta = \pi/n$ is mapped to the upper half plane by the transformation $w = z^n$ if $n \geq 1/2$.

Conformal Mapping

Let C_1 and C_2 be two curves in the z plane. Suppose that a given transformation w maps these curves to curves C_1' and C_2' in the w plane. Let θ be the angle between the curves C_1 and C_2 and ϕ be the angle between curves C_1' and C_2'. If $\theta = \phi$ in both magnitude and sense, we say that the mapping is *conformal*. Put another way, a conformal mapping preserves angles. There is an important theorem related to conformal mappings.

Suppose that $f(z)$ is analytic and that $f'(z) \neq 0$ in some region R of the complex plane. It follows that the mapping $w = f(z)$ is conformal at all points of R.

The Mapping 1/z

Consider the transformation $w = 1/z$. Notice that we can write

$$w = \frac{1}{z} = \frac{\bar{z}}{z\bar{z}} = \frac{x - iy}{(x+iy)(x-iy)} = \frac{x}{x^2 + y^2} - i\frac{y}{x^2 + y^2}$$

By inverting the transformation, it is easy to see that the coordinates in the z plane are related to coordinates in the w plane in the same way:

$$x = \frac{u}{u^2 + v^2} \qquad y = -\frac{v}{u^2 + v^2}$$

Therefore the mapping $w = 1/z$

- Transforms lines in the z plane to lines in the w plane
- Transforms circles in the z plane to circles in the w plane

In particular

- A circle that does not pass through the origin in the z plane is transformed into a circle not passing through the origin in the w plane.
- A circle passing through the origin in the z plane is transformed into a line that does not pass through the origin in the w plane.
- A line not passing through the origin in the z plane is transformed into a circle through the origin in the w plane.
- A line through the origin in the z plane is transformed into a line through the origin in the w plane.

Let's try to understand how $w = 1/z$ maps lines into lines. A line in the complex plane that passes through the origin is a set of points of the form

$$z = re^{ia}$$

where a is some fixed angle. Under the mapping $w = 1/z$, we obtain a set of points:

$$w = \frac{1}{z} = \frac{1}{r} e^{-ia}$$

This is another line that passes through the origin.

An important mapping $w = 1/z$ transforms a disk in the z plane into the exterior of the disk in the w plane. Consider the disk shown in Fig. 9.10. The mapping $w = 1/z$ maps this to the *exterior* of the circle of radius $1/r$. This is illustrated in Fig. 9.11.

The mapping $w = 1/z$ takes the point $z = 0$ to $z = \infty$, and takes $z = \infty$ to $z = 0$.

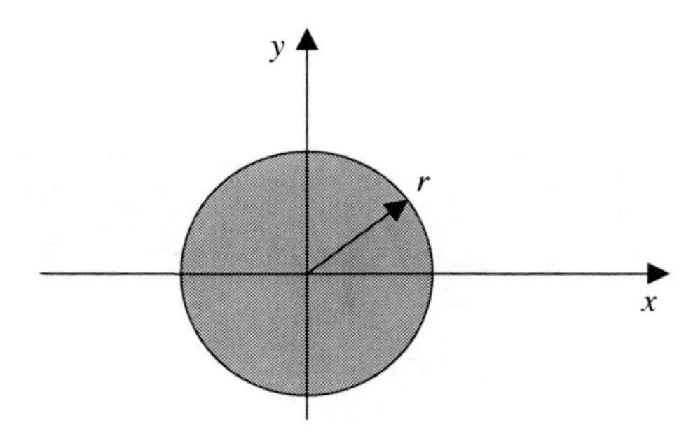

Figure 9.10 A disk of radius r in the z plane.

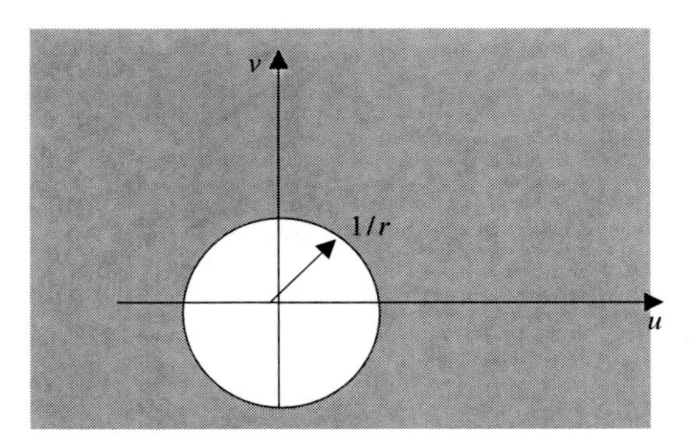

Figure 9.11 The transformation $w = 1/z$ has mapped the disk in Fig. 9.9 to the entire w plane *minus the region covered by the disk of radius* $1/r$.

Mapping of Infinite Strips

There are several important transformations that can be applied to infinite strips to map them to the upper half of the w plane. Consider a strip of height a in the y direction that extends to $\pm\infty$ along the x axis. This is illustrated in Fig. 9.12. When figuring out how a transformation will work out, we pick out a few key points. These are denoted by A-F in the figure.

The exponential function sends horizontal lines in the z plane into *rays* in the w plane. That is, consider the transformation

$$w = e^z$$

This maps the lines as shown in Fig. 9.13.

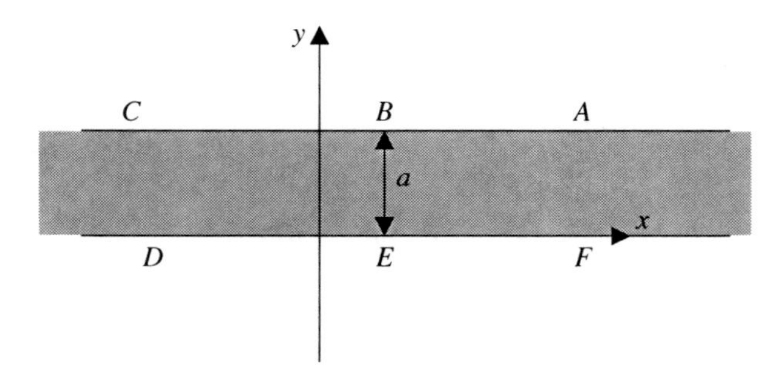

Figure 9.12 An infinite strip in the z plane.

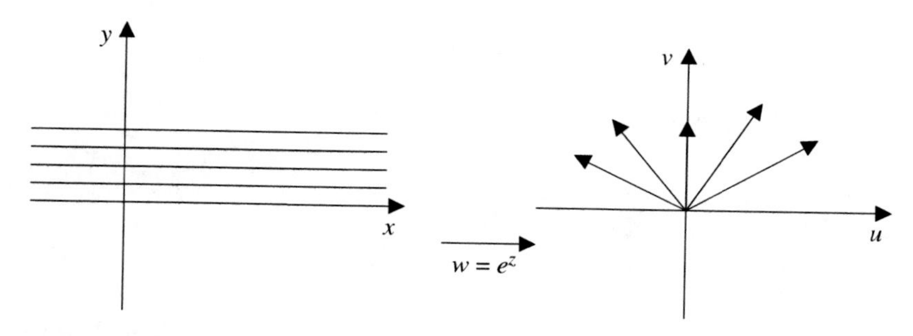

Figure 9.13 The exponential function maps horizontal lines to rays.

If we apply the transformation

$$w = e^{\pi z/a} \tag{9.2}$$

to the infinite strip shown in Fig. 9.12, the result is a mapping to the upper half of the w plane, shown in Fig. 9.14. The points A, B, C, D, E, and F map to the points A', B', C', D', E', and F', respectively.

Now consider a vertical strip, as shown in Fig. 9.15. We can map this to the upper half plane of Fig. 9.14 using the transformation

$$w = \sin\frac{\pi z}{a} \tag{9.3}$$

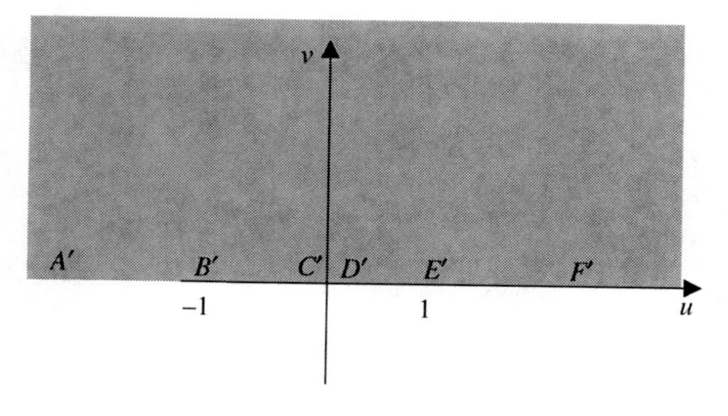

Figure 9.14 A mapping $w = e^{\pi z/a}$ to the infinite strip shown in Fig. 9.11 maps it to the upper half plane.

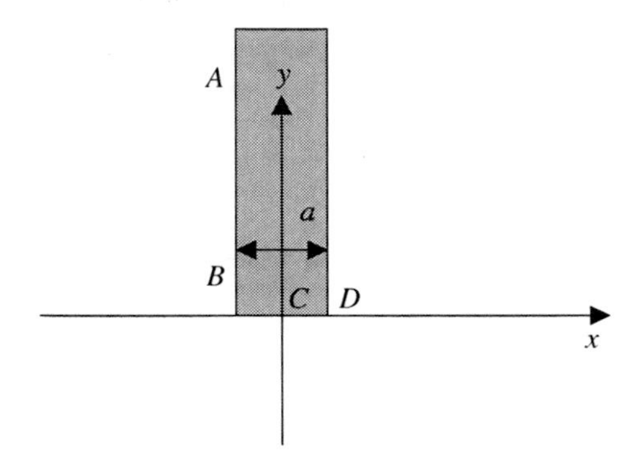

Figure 9.15 To map a vertical strip to the upper half plane, we utilized the transformation $w = \sin \pi z / a$. It maps to the region shown in Fig. 9.16.

Rules of Thumb

Here are a few basic rules of thumb to consider when doing transformations. They can be illustrated considering the square region shown in Fig. 9.6. In that section, we saw how to shift the position of the square by adding a constant, that is, we wrote down a linear transformation of the form $w = z + a$. Let's review the other types of transformations that are possible. A transformation of the type

$$w = az$$

will *expand* the region if $|a| > 1$ and will *shrink* the region if $|a| < 1$. Consider the first case with $w = 2z$. This expands the square region from $0 \leq x \leq 1, 0 \leq y \leq 1$ to $0 \leq x \leq 2, 0 \leq y \leq 2$. This is illustrated in Fig. 9.17.

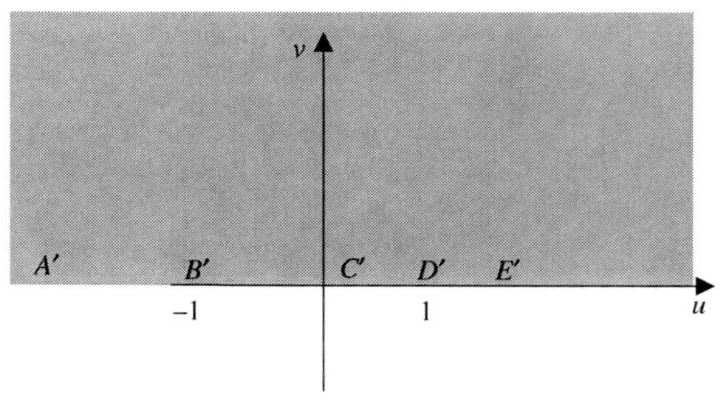

Figure 9.16 The transformation $w = \sin(\pi z / a)$ maps the region in Fig. 9.15 to the region in Fig. 9.16 with corresponding points indicated.

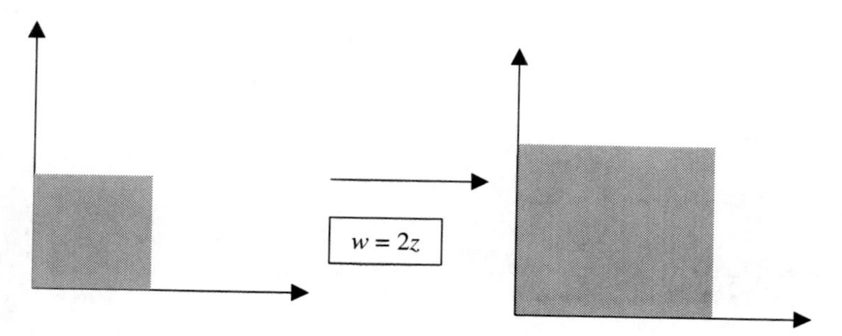

Figure 9.17 For review, a transformation of the form $w = az$ expands the region of interest if $|a| > 1$.

Now suppose that $w = (1/2)z$. This shrinks the square, as shown in Fig. 9.18. To rotate the region by an angle ϕ, we use a transformation of the form

$$w = e^{i\phi} z \tag{9.4}$$

For our square, this rotates the square by ϕ in the counterclockwise direction assuming that $\phi > 0$. This is illustrated in Fig. 9.19.

Möbius Transformations

In this section, we consider a transformation of the type:

$$Tz = \frac{az+b}{cz+d} \qquad ad - bc \neq 0 \tag{9.5}$$

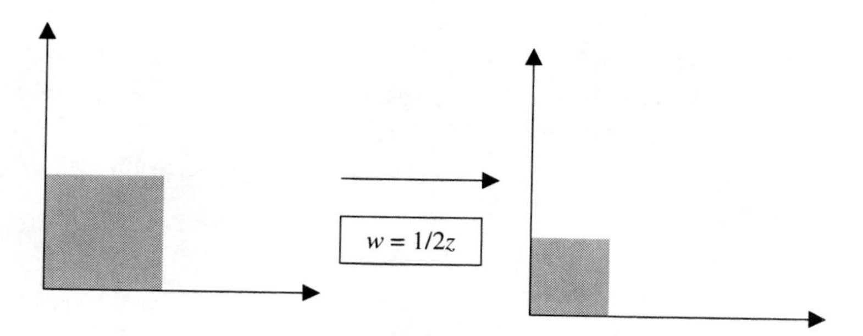

Figure 9.18 We shrink the square by the transformation $w = az$ when $|a| < 1$.

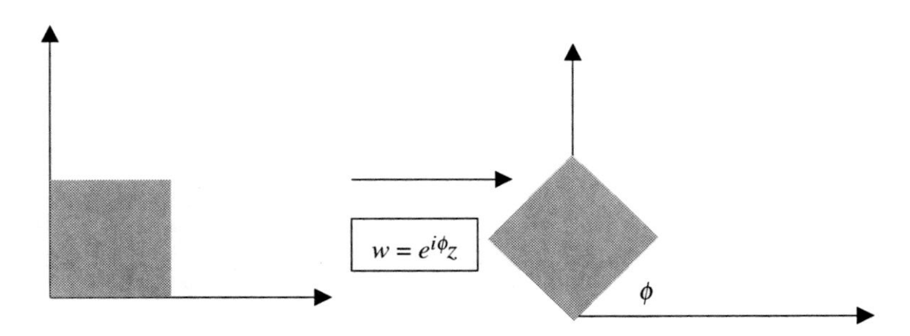

Figure 9.19 A rotation is implemented with a transformation of the form $w = e^{i\phi}z$.

This type of transformation goes under the various names *bilinear transformation, fractional transformation,* or *Möbius transformation.* The transformation shown in Eq. (9.5) is actually a composition of three different transformations. These are

- *Dilation*, which can be written as the linear transformation az.
- *Translation*, which is written as $z+b$.
- *Reciprocation*, which is the transformation $1/z$.

The requirement that $ad-bc \neq 0$ is based on the following. The derivative of Eq. (9.5) is given by

$$(Tz)' = \frac{a(cz+d)-c(az+b)}{(cz+d)^2}$$

Evaluating this at $z=0$ we have

$$(Tz)'(0) = \frac{ad-bc}{(d)^2}$$

This tells us that the transformation in Eq. (9.5) will be a *constant* unless $ad-bc \neq 0$. A transformation of the type in Eq. (9.5) maps circles in the z plane to circles in the w plane. Straight lines are also mapped into straight lines.

Now suppose that $z_0, z_1, z_2,$ and z_3 are four distinct points in the complex plane. The *cross ratio* is given by

$$\frac{(z_3-z_0)(z_1-z_2)}{(z_1-z_0)(z_3-z_2)} \tag{9.6}$$

The cross ratio is invariant under a Möbius transformation. That is if $z_j \to w_j$ under a Möbius transformation, then

$$\frac{(z_3 - z_0)(z_1 - z_2)}{(z_1 - z_0)(z_3 - z_2)} \to \frac{(w_3 - w_0)(w_1 - w_2)}{(w_1 - w_0)(w_3 - w_2)}$$

There are a few Möbius transformations of interest. Let a be a complex number with $|a| < 1$ and suppose that $|k| = 1$. Then

$$w = k\frac{z - a}{1 - \bar{a}z} \tag{9.7}$$

maps the unit disk from the z plane to the unit disk in the w plane. Now let a be a complex number with the requirement that $\text{Im}(a) > 0$. The transformation

$$w = k\frac{z - a}{z - \bar{a}} \tag{9.8}$$

maps the upper half of the z plane to the unit disk in the w plane. Notice that when z is purely real, $|w| = |k| = 1$.

EXAMPLE 9.4

Consider a disk of radius $r = 2$ centered at the point $z = -1 + i$. Find a transformation that will take this to the entire complex plane with a hole of radius 1/2 centered at the origin.

SOLUTION

Since these transformations are linear, we can do this by taking multiple transformations in succession. First we illustrate what we're starting with, a disk of radius $r = 2$ centered at the point $z = -1 - i$. This is shown in Fig. 9.20.

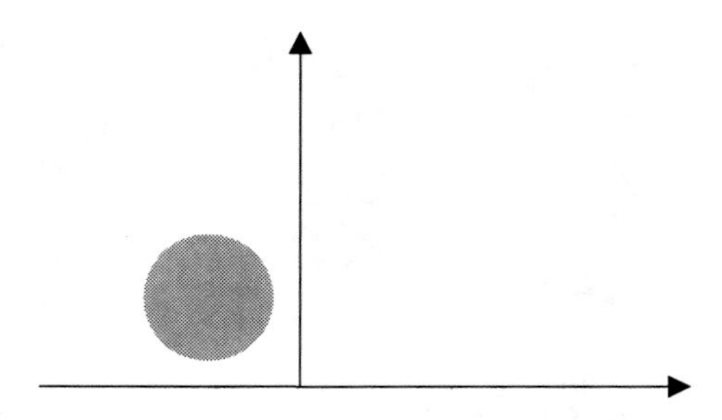

Figure 9.20 In Example 9.4, we start with a region defined by a disk centered at $z = -1 + i$.

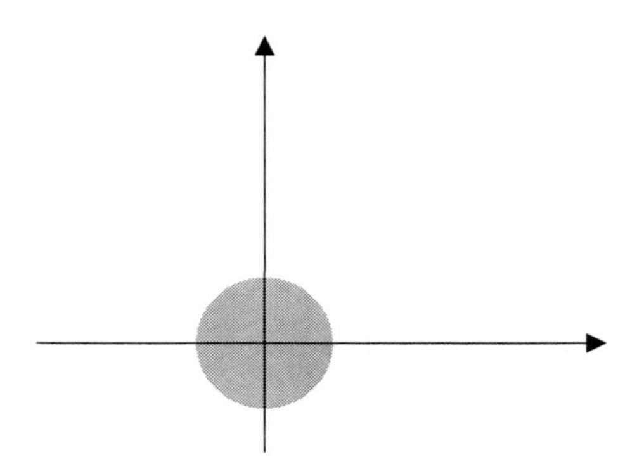

Figure 9.21 A disk at the origin is obtained from the disk shown in Fig. 9.20 via the transformation $Z = z - z_0 = z + 1 - i$. We denote this the Z plane.

The first step is to move the disk to the origin. We do this using

$$Z = z - z_0 = z + 1 - i$$

The result is the disk shown in Fig. 9.21.

Now we want to transform the disk shown in Fig. 9.21 so that the region of definition is the entire complex plane minus a hole where the disk was. We do this using an inverse transformation:

$$w = \frac{1}{Z}$$

The result is shown in Fig. 9.22.

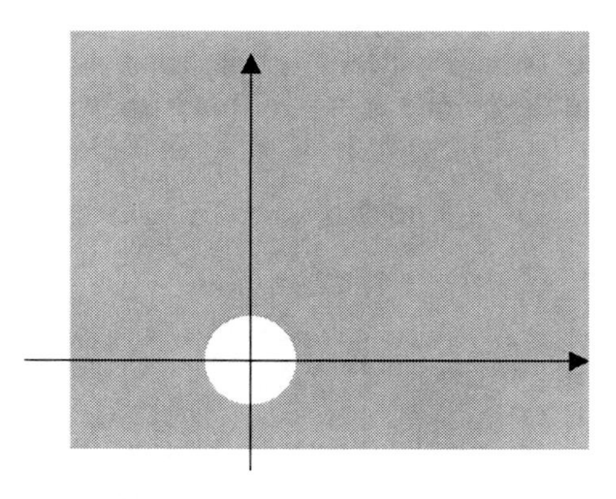

Figure 9.22 The transformation $1/Z$ changes the region to the entire complex plane with a hole punched out in the middle. The radius of the hole is $1/r$ if the radius of the disk we started with was $Z = re^{i\theta}$. In our example, $r = 2$ so the hole here has a radius $\rho = 1/2$.

The complete transformation in this example can be written as

$$w = \frac{1}{z+1-i}$$

This is a Möbius transformation as in Eq. (9.5) with $a = 0,\ b = 1, c = 1,$ and $d = -1 + i$.

EXAMPLE 9.5

Construct a Möbius transformation that maps the unit disk to the left half plane $\text{Re}(z) < 0$ and one that maps the unit disk to the right half plane $\text{Re}(z) > 0$.

SOLUTION

The first transformation we want to consider is illustrated in Fig. 9.23.

First we consider the boundary of the disk, which is the unit circle, that is the set of points $|z| = 1$. For the transformation shown in Fig. 9.23 to work, we must map the points on the unit circle to the imaginary axis. In the form of a Möbius transformation, the mapping will be of the form

$$Tz = \frac{az+b}{cz+d}$$

This transformation has a pole located at the point $z = -d/c$. We are free to pick a point on the unit circle to map to the pole, so we choose $z = 1$. With this choice we have the freedom to fix c and d, so we choose $c = 1,\ d = -1$. So

$$Tz = \frac{az+b}{z-1}$$

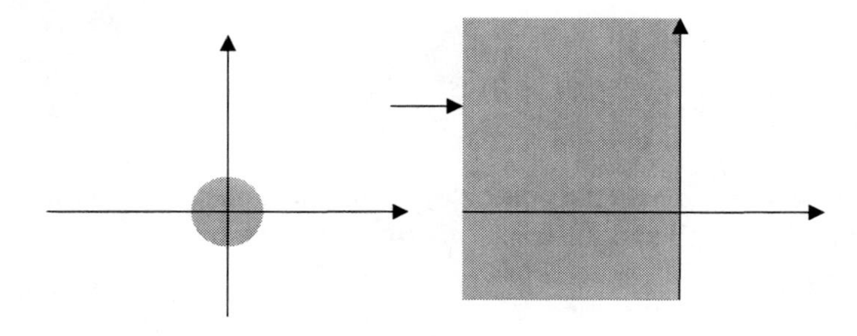

Figure 9.23 We want to map the unit disk to the left half plane.

Second, we need to pick a point z on the unit circle such that $Tz = 0$. We have already used the point $z = 1$, so we choose $z = -1$. This forces us to take $a = b$ since $Tz = 0$ when $z = -1$. We can choose $a = 1$ giving us the transformation

$$Tz = \frac{z+1}{z-1}$$

Now we can see how the transformation maps the rest of the unit disk. The simplest point to check is the point $z = 0$ at the center of the disk. We have

$$Tz(0) = \frac{0+1}{0-1} = -1$$

So the transformation maps the center of the disk $z = 0$ to the point $z = -1$ which is in the left half plane. Hence this is the transformation that we want.

To transform the unit disk to the right half plane instead, it turns out we are almost there. All we have to do is rotate the transformed region shown in Fig. 9.22. The angle that is required is π, and a rotation is implemented by multiplication by $e^{i\theta}$. So the transformation that takes the unit disk to the right half plane is given by

$$Tz = e^{i\pi} \frac{z+1}{z-1} = -\left(\frac{z+1}{z-1}\right)$$

EXAMPLE 9.6

Consider a mapping that will transform the unit disk into the upper half plane.

SOLUTION

We again seek a transformation of the form

$$Tz = \frac{az+b}{cz+d}$$

This time we choose to map the point $z = -1$ onto the pole at $z = -d/c$. If we choose $c = 1$, then $d = 1$ as well and the transformation is given by

$$Tz = \frac{az+b}{z+1}$$

Following the last example, now we need to pick a point z on the unit circle such that $Tz = 0$. We have already used the point $z = -1$, so we choose $z = +1$. This forces us to take $a = -b$ since $Tz = 0$ when $z = 1$. We can choose $b = 1$ giving us the transformation

$$Tz = \frac{1 - z}{z + 1}$$

Now we check how this transformation maps the point $z = 0$ at the center of the disk. We find

$$Tz(0) = \frac{1 - 0}{0 + 1} = +1$$

This is a point in the right half plane. So the transformation does not map to the upper half plane. We can make it do so with another rotation, given by $e^{i\pi/2} = i$. Hence, the transformation that will map the unit circle to the upper half plane is

$$Tz = i\frac{1 - z}{z + 1}$$

Notice that it takes the point $z = 0 \rightarrow i$, which is in the upper half plane.

Fixed Points

The *fixed points* of a transformation are those for which

$$Tz = \frac{az + b}{cz + d} = z \tag{9.9}$$

A fixed point is one that is left *invariant* by a transformation.

EXAMPLE 9.7

Find the fixed points of

$$Tz = \frac{z - 1}{z + 4}$$

SOLUTION

The fixed points are those for which

$$\frac{z-1}{z+4} = z$$
$$\Rightarrow z^2 + 3z + 1 = 0$$

There are two fixed points:

$$z = -\frac{3}{2} \pm \frac{\sqrt{5}}{2}$$

Summary

In this chapter, we introduced the notion of a transformation, which allows us to transform a region in the complex plane into a different region. This is a useful technique for solving differential equations, among other applications.

Quiz

1. Consider a horizontal line $y = a$ in the z plane and let $w = z^2$. What kind of curve results in the w plane?
2. Let $x = a$ be a vertical line in the z plane, and let $w = e^z$. What kind of curve results in the w plane?
3. Construct a Möbius transformation that maps the upper half of the z plane to the unit disk in the w plane with $z = i \to w = 0$ and the point at infinity is mapped to $w = -1$.
4. Find the fixed points of the transformation $Tz = \frac{z+1}{z-1}$.
5. Find a Möbius transformation that maps the points $z = \{-1, 0, 1\}$ onto $w = \{-i, 1, i\}$.
6. Find a Möbius transformation that maps the upper half plane ($y > 0$) to the half plane $v > 0$ and the x axis to the u axis.

CHAPTER 10

The Schwarz-Christoffel Transformation

The *Schwarz-Christoffel transformation* is a transformation that maps a simple closed polygon to the upper half plane. The transformation can be used in applications such as fluid dynamics and electrostatics. In this chapter, we will introduce some basics about the transformation.

The Riemann Mapping Theorem

The *Riemann mapping theorem* establishes the existence of a transformation that will map a region R of the z plane to a region R' of the w plane. Let $w = f(z)$ be an analytic function in R and let R be enclosed by a simple closed curve C.

Suppose that the region R' is the unit disk at the origin bounded by the unit circle C', that is a circle with radius $|w| = 1$.

The Riemann mapping theorem says that the function $w = f(z)$ exists, that it maps each point of the region R into a point in R', and it maps each point on C to a point on C'. Furthermore this mapping is one to one. There are three arbitrary real constants associated with the mapping $w = f(z)$. To find them, we establish a correspondence between the origin of the w plane and a point belonging to R and between a point on C' and a point on C.

If $z_0 \in R$ with $f(z_0) = 0$ and $f'(z_0) > 0$ then the mapping $w = f(z)$ is unique.

The Schwarz-Christoffel Transformation

The Schwarz-Christoffel transformation maps

- The interior of a polygon to the upper half plane
- The boundary of the polygon to the real axis

Here we give a heuristic explanation of the transform (not a formal derivation) and state the result. Our discussion is out of Levinson and Redheffer (see the bibliography list at the end of the book).

Consider the polygon shown in Fig. 10.1. In what follows, we assume the polygon is in the w plane and that it is mapped to the upper half of the z plane.

As noted in the figure, the curve enclosing the region is traversed in the positive or counterclockwise sense. We define the interior angles at the vertices by

$$\alpha_1 \pi, \alpha_2 \pi, \ldots, \alpha_n \pi$$

Now we assume the existence of a function $w = f(z)$ that maps the interior of the polygon to the upper half plane (the existence of the function is implied by the Riemann mapping theorem). If this mapping is one-to-one and conformal (angle preserving) then it follows that $f(z)$ is analytic for $y > 0$ and continuous for $y \geq 0$.

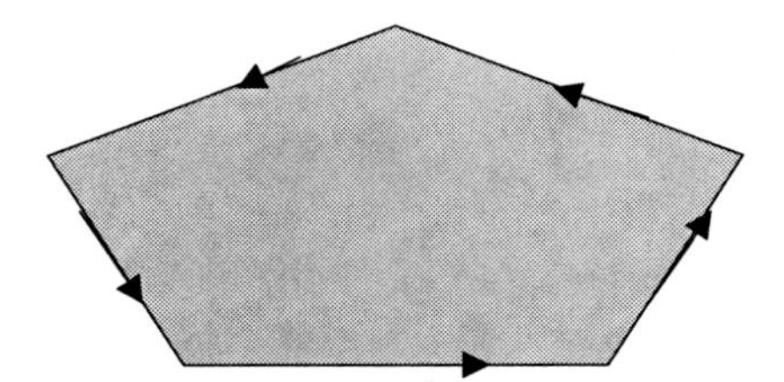

Figure 10.1 A polygon to be mapped to the upper half plane.

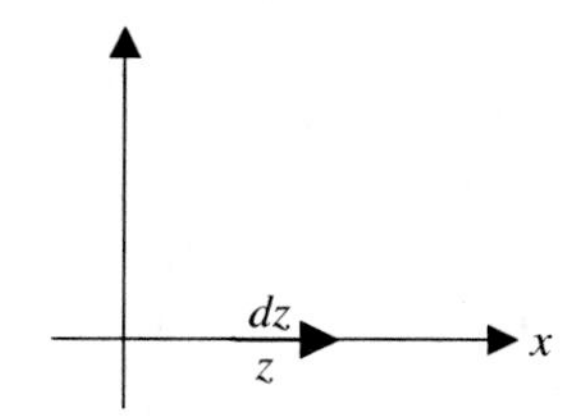

Figure 10.2 dz is a vector on the x axis of the z plane.

We can also assume the existence of an inverse mapping, which we denote by $z = g(w)$. The inverse mapping is analytic on the polygon's interior. Moreover, it is continuous in the interior of the polygon and on it's boundary.

The boundary of the polygon is mapped onto the real axis. Suppose that the vertices of the polygon are mapped onto the points of the real x axis denoted by

$$x_1, x_2, \ldots, x_n$$

Now, since $w = f(z)$, it follows that

$$dw = f'(z)\,dz \tag{10.1}$$

Next we assume that the mapping and its inverse $w = f(z), z = g(w)$ are analytic on the sides of the polygon in addition to its interior. Picking a point w on the polygon which is not a vertex, the image of w is a point z. The point dz is a positive vector on the real axis x. This is shown in Fig. 10.2.

It should follow that dw is a vector on the edge of the polygon pointing in the positive sense (in the counterclockwise direction). This is shown in Fig. 10.3.

The arguments of f' and w' are then related in the following way:

$$\arg[f'(z)] = \arg\left(\frac{dw}{dz}\right) \tag{10.2}$$

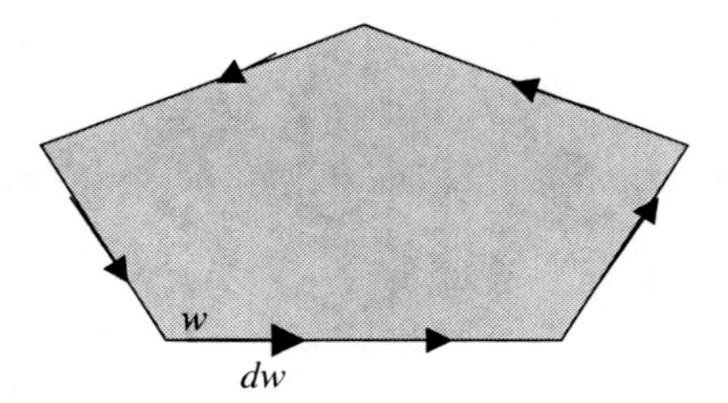

Figure 10.3 If we take dz to be a positive pointing vector on the real axis, then dw is a vector on the edge of the polygon pointing in the positive sense.

Now imagine moving the point w around the polygon and moving the image point z in the positive direction along the x axis. Each time a point w moves past a vertex of angle $\alpha_1\pi$, the argument changes as $\pi(1-\alpha_1)$. But $\arg dz = 0$. Therefore, to the left of the point x_1 $\arg[f'(z)]$ is a constant, but at the point x_1 it will change by $\pi(1-\alpha_1)$ to maintain Eq. (10.2).

As z moves from $x < x_1$ to $x > x_1$, $\arg[(z-x_1)]$ decreases from π to 0. This means that the argument of $(z-x_1)^{\alpha_1-1}$ is changed by $\pi(1-\alpha_1)$. This change will occur at every vertex. So we choose

$$f'(z) = A(z-x_1)^{\alpha_1/\pi-1}(z-x_2)^{\alpha_2/\pi-1}\cdots(z-x_n)^{\alpha_n/\pi-1} \tag{10.3}$$

Then, using Eq. (10.1)

$$\frac{dw}{dz} = A(z-x_1)^{\alpha_1/\pi-1}(z-x_2)^{\alpha_2/\pi-1}\cdots(z-x_n)^{\alpha_n/\pi-1} \tag{10.4}$$

Integrating, we obtain the form $w = f(z)$:

$$w = A\int(z-x_1)^{\alpha_1/\pi-1}(z-x_2)^{\alpha_2/\pi-1}\cdots(z-x_n)^{\alpha_n/\pi-1}dz + B \tag{10.5}$$

The constants A and B are in general complex that indicate the orientation, size, and location of the pentagon in the w plane. To obtain the mapping in Eq. (10.5), three points out of $x_1, x_2, \ldots, x_n$ can be chosen. If a point x_j is taken at infinity then the factor $(z-x_j)^{\alpha_j/\pi-1}$ is not included in Eq. (10.5).

EXAMPLE 10.1

Find the image in the upper half plane using the transformation

$$w = \int_0^z \frac{1}{\sqrt{(1-t^2)(1-k^2t^2)}}dt$$

when $0 < k < 1$.

SOLUTION

Looking at Eq. (10.5), we see that this mapping is a Schwarz-Christoffel transformation. Let's rewrite the integral in a more suggestive form:

$$w = \int_0^z (1-t^2)^{-1/2}(1-k^2t^2)^{-1/2}dt$$

Now the transformation looks like Eq. (10.5). To find the vertices of the polygon, we look for the zeros of the integrand. First consider $1-t^2 = 0$, which tells us that $z = \pm 1$. Next we have $1-k^2t^2 = 0$ from which it follows that $z = \pm(1/k)$. Since each term in the Schwarz-Christoffel transformation is of the form $(z-x_j)^{\alpha_j/\pi-1}$

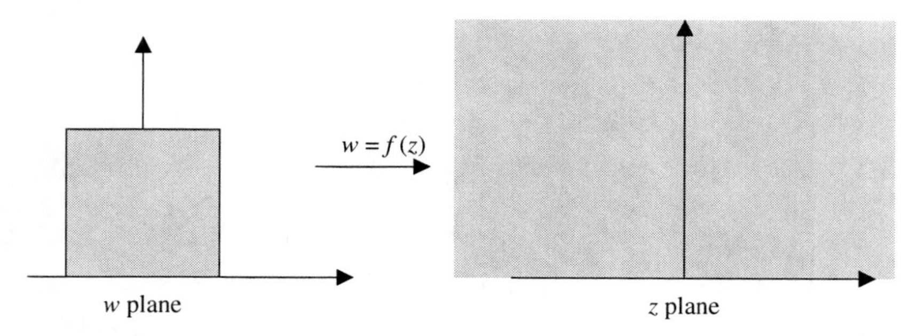

Figure 10.4 The transformation in Example 10.1 maps a rectangle to the upper half plane.

and the exponents in this example are $\alpha_j/\pi - 1 = -1/2$, it follows that $\alpha_1 = \alpha_2 = \pi/2$, or each angle is increased by $\pi/2$ at each vertex. The polygon described by this is a rectangle. The height of the rectangle is found from

$$h = \int_1^{1/k} (1-t^2)^{-1/2}(1-k^2t^2)^{-1/2}dt$$

The width of the rectangle is

$$W = 2\int_0^1 (1-t^2)^{-1/2}(1-k^2t^2)^{-1/2}dt$$

Looking at the definition of the transformation, $w(0) = \int_0^0 \left[1/\sqrt{(1-t^2)(1-k^2t^2)}\right]dt = 0$, so the origin of the w plane is mapped to the origin of the z plane. The vertices of the polygon are located at $(-W,0), (W,0), (W,ih)$, and $(-W,ih)$. The transformation is illustrated in Fig. 10.4.

Summary

In this chapter, we wrapped up the discussion of mappings or transformations begun in Chap. 9. First we stated the Riemann mapping theorem, which guarantees the existence of a mapping between a region of the w plane and a region of the z plane. Next we introduced the Schwarz-Christoffel transformation, which allows us to map a polygon to the upper half plane.

Quiz

1. What type of region does the transformation $w = A\int \frac{dt}{\sqrt{1-t^2}} + B$ map to the upper half plane?

The Gamma and Zeta Functions

In this chapter, we review two important functions related to complex analysis, the *gamma* and *zeta* functions.

The Gamma Function

The gamma function can be defined in terms of complex variable z provided that $\mathrm{Re}(z) > 0$ as follows:

$$\Gamma(z) = \int_0^\infty t^{z-1} e^{-t} dt \tag{11.1}$$

We can show that Eq. (11.1) is convergent in the right-hand plane by examining its behavior at $t = 0$ and as $t \to \infty$. Using $|z| \le \operatorname{Re} z$ we have

$$\left|t^{z-1}e^{-t}\right| \le t^{\operatorname{Re} z-1}$$

This tells us that Eq. (11.1) is finite at $t = 0$. As $t \to \infty$, note that we have

$$\left|t^{z-1}e^{-t}\right| \le t^{\operatorname{Re} z-1}e^{-t} \le e^{-t/2}$$

This shows that the integral is convergent for large t. If n is a positive integer, then

$$\Gamma(n+1) = n(n-1)\cdots 1 = n! \tag{11.2}$$

This follows from the *recursion relation* for the gamma function, which holds for any z (not just integers):

$$\Gamma(z+1) = z\Gamma(z) \tag{11.3}$$

EXAMPLE 11.1

Prove the recursion relation in Eq. (11.3) for the gamma function.

SOLUTION

This can be done using the definition in Eq. (11.1). We calculate $\Gamma(z+1)$:

$$\Gamma(z+1) = \int_0^\infty t^z e^{-t}\,dt$$

Now notice what happens if we take the derivative of the integrand:

$$\frac{d}{dt}(t^z e^{-t}) = z\,t^{z-1}e^{-t} - t^z e^{-t}$$

It follows that

$$t^z e^{-t} = z\,t^{z-1}e^{-t} - \frac{d}{dt}(t^z e^{-t})$$

So the integral can be written as

$$\begin{aligned}\Gamma(z+1) &= \int_0^\infty t^z e^{-t} dt \\ &= \int_0^\infty \left[z t^{z-1} e^{-t} - \frac{d}{dt}(t^z e^{-t}) \right] dt \\ &= \int_0^\infty z t^{z-1} e^{-t} dt - \int_0^\infty \frac{d}{dt}(t^z e^{-t}) dt\end{aligned}$$

Looking at the second term, we have

$$\int_0^\infty \frac{d}{dt}(t^z e^{-t}) dt = (t^z e^{-t}) \Big|_{t=0}^\infty = \lim_{t \to \infty} t^z e^{-t} - \lim_{t \to 0} t^z e^{-t} = 0 - 0 = 0$$

Therefore

$$\begin{aligned}\Gamma(z+1) &= \int_0^\infty t^z e^{-t} dt \\ &= \int_0^\infty z t^{z-1} e^{-t} dt \\ &= z \int_0^\infty t^{z-1} e^{-t} dt \\ \Rightarrow \Gamma(z+1) &= z\Gamma(z)\end{aligned}$$

EXAMPLE 11.2

Show that an alternative definition of the gamma function is given by

$$\Gamma(z) = 2\int_0^\infty x^{2z-1} e^{-x^2} dx \tag{11.4}$$

SOLUTION

This is a simple substitution problem. We start with Eq. (11.1) and choose

$$t = x^2$$

It follows that

$$dt = 2x dx$$

Hence

$$\begin{aligned}\Gamma(z) &= \int_0^\infty t^{z-1}e^{-t}\,dt \\ &= \int_0^\infty (x^2)^{z-1}e^{-x^2}\,2x\,dx \\ &= 2\int_0^\infty x^{2z-2}e^{-x^2}\,x\,dx \\ &= 2\int_0^\infty x^{2z-1}e^{-x^2}\,dx\end{aligned}$$

This works provided that $\mathrm{Re}(z) > 0$.

EXAMPLE 11.3
Show that $\int_0^1 x^n \ln x\, dx = -1/(1+n)^2$, if $n > -1$.

SOLUTION
We begin by making the substitution

$$x = e^u$$

Then of course:

$$dx = e^u du$$

Now the integral can be written in the following way:

$$\begin{aligned}\int_0^1 x^n \ln x\, dx &= \int_\infty^0 (e^u)^n \ln(e^u) e^u du \\ &= -\int_0^\infty (e^u)^n \ln(e^u) e^u du \\ &= -\int_0^\infty u e^{u(1+n)} du\end{aligned}$$

In the first step, we used $x = e^u$ to note that, when $x = 0, \Rightarrow u = \infty$, and when $x = 1$, $u = 0$ To see how this works note that, $\ln x = \ln(e^u) = u, \therefore x = 0 \Rightarrow u = \ln 0 = \infty$. Now

we can do another substitution. This time we let $-t = u(1+n)$. Then $dt = -(1+n)du$. And so

$$\int_0^1 x^n \ln x\,dx = -\int_0^\infty ue^{u(1+n)}du$$
$$= -\int_0^\infty e^{-t}\left(\frac{-t}{1+n}\right)\frac{dt}{-(1+n)}$$
$$= -\int_0^\infty e^{-t}\left[\frac{t}{(1+n)^2}\right]dt$$
$$= -\frac{1}{(n+1)^2}\int_0^\infty e^{-t}t\,dt$$

But, using Eq. (11.1) together with Eq. (11.2)

$$\int_0^\infty e^{-t}t\,dt = \Gamma(2) = 1! = 1$$

Therefore

$$\int_0^1 x^n \ln x\,dx = -\frac{1}{(1+n)^2}$$

EVALUATING $\Gamma(z)$ WHEN $0 < Z < 1$

It is given as a definition in most texts that

$$\Gamma\left(\frac{1}{2}\right) = \sqrt{\pi} \tag{11.5}$$

Using the recursion formula in Eq. (11.3), it is possible to evaluate the gamma function for $0 < z < 1$ if $\text{Re}(z) > 0$. This result is established in the next example.

EXAMPLE 11.4

Show that $\Gamma(z)\Gamma(1-z) = \pi/\sin \pi z$, and hence that $\Gamma(1/2) = \sqrt{\pi}$.

SOLUTION

In Example 11.2, we showed that

$$\Gamma(z) = 2\int_0^\infty x^{2z-1}e^{-x^2}dx$$

It follows that (considering real z such that $0 < z < 1$):

$$\Gamma(n)\Gamma(1-n) = \left\{2\int_0^\infty x^{2n-1}e^{-x^2}dx\right\}\left\{2\int_0^\infty y^{1-2n}e^{-y^2}dy\right\}$$
$$= 4\int_0^\infty\int_0^\infty x^{2n-1}y^{1-2n}e^{-(x^2+y^2)}dx\,dy$$

Now rewrite the integral in terms of polar coordinates $x = r\cos\theta$, $y = r\sin\theta$ to give

$$\Gamma(n)\Gamma(1-n) = 4\int_0^{\pi/2}\int_0^\infty \tan^{1-2n}\theta\, re^{-r^2}drd\theta$$

Integration over r can be done readily yielding

$$\Gamma(n)\Gamma(1-n) = 2\int_0^{\pi/2}\tan^{1-2n}\theta\, d\theta \tag{11.6}$$

In order to calculate Eq. (11.6), we will have to take a major aside. We will show how to calculate the integral

$$\int_0^\infty \frac{x^{p-1}}{1+x}dx$$

using residue theory. This can be done by calculating the contour integral

$$\oint \frac{z^{p-1}}{1+z}dz$$

The point $z = 0$ is a branch point and the point $z = -1$ is a simple pole. We can deal with the branch point by considering the contour shown in Fig. 11.1.

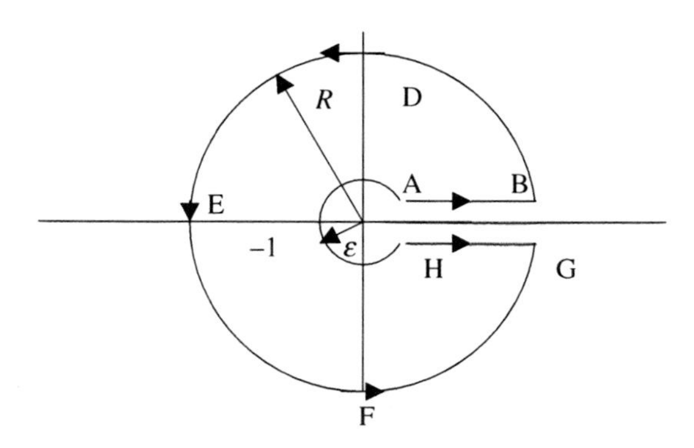

Figure 11.1 The contour used to evaluate $\oint[z^{p-1}/(1+z)]dz$ has an inner circle of radius ε and an outer circle of radius R. It encloses the singularity at $z = -1$ while avoiding the branch point at $z = 0$.

Now, recalling that $-1 = e^{i\pi}$ we find that the residue corresponding to the singularity at $z = -1$ is given by

$$\lim_{z\to -1}(1+z)\frac{z^{p-1}}{1+z} = \lim_{z\to e^{i\pi}} z^{p-1} = e^{i(p-1)\pi}$$

Hence by the Cauchy residue theorem

$$\oint \frac{z^{p-1}}{1+z}dz = 2\pi i e^{i(p-1)\pi}$$

This means that integrating around the contour we will have

$$\int_{AB} + \int_{BDEFG} + \int_{GH} + \int_{HJA} = 2\pi i e^{i(p-1)\pi}$$

On the large exterior circle we have $z = R\mathrm{e}^{i\theta}$ while on the small interior circle we have $z = \varepsilon e^{i\theta}$. That is the integral can be written as

$$\int_{\varepsilon}^{R}\frac{x^{p-1}}{1+x}dx + \int_{0}^{2\pi}\frac{(Re^{i\theta})^{p-1}iRe^{i\theta}d\theta}{1+Re^{i\theta}} + \int_{R}^{\varepsilon}\frac{(xe^{2\pi i})^{p-1}dx}{1+xe^{2\pi i}} + \int_{2\pi}^{0}\frac{(\varepsilon e^{i\theta})^{p-1}i\varepsilon e^{i\theta}d\theta}{1+\varepsilon e^{i\theta}}$$

$$= 2\pi i e^{i(p-1)\pi}$$

Now take the limit as $\varepsilon \to 0$ and $R \to \infty$. Then

$$\int_{BDEFG} = \int_{HJA} = 0$$

Giving

$$\int_{0}^{\infty}\frac{x^{p-1}}{1+x}dx - \int_{0}^{\infty}\frac{(xe^{2\pi i})^{p-1}dx}{1+xe^{2\pi i}} = 2\pi i e^{i(p-1)\pi}$$

$$\Rightarrow (1-e^{2\pi(p-1)i})\int_{0}^{\infty}\frac{x^{p-1}}{1+x}dx = 2\pi i e^{i(p-1)\pi}$$

This allows us to write

$$\int_{0}^{\infty}\frac{x^{p-1}}{1+x}dx = \frac{2\pi i\, e^{\pi(p-1)i}}{1-e^{2\pi(p-1)i}} = \pi\frac{2i}{e^{p\pi i}-e^{-p\pi i}} = \frac{\pi}{\sin p\pi}$$

After this long detour, we have calculated Eq. (11.6):

$$\Gamma(n)\Gamma(1-n) = 2\int_0^{\pi/2} \tan^{1-2n}\theta\, d\theta = \frac{\pi}{\sin n\pi} \tag{11.7}$$

Setting $n = 1/2$ it follows that

$$\left[\Gamma\left(\frac{1}{2}\right)\right]^2 = \frac{\pi}{\sin(\pi/2)} = \pi$$

$$\Rightarrow \Gamma\left(\frac{1}{2}\right) = \sqrt{\pi}$$

Note that while we calculated Eq. (11.7) for real z such that $0 < z < 1$, the result can be extended using analytic continuation.

EXAMPLE 11.5

Using the gamma function, show that $\int_0^\infty e^{-x^4} dx$.

SOLUTION

This integral can be written as a gamma function by using a substitution. We take

$$u = x^2 \qquad \Rightarrow du = 2xdx$$

That is:

$$x = \sqrt{u} \qquad \Rightarrow dx = \frac{du}{2\sqrt{u}}$$

Substituting we find

$$\int_0^\infty e^{-x^4} dx = \int_0^\infty e^{-u^2} u^{-1/2} \frac{du}{2}$$

$$= \frac{1}{2}\int_0^\infty e^{-u^2} u^{-1/2} du$$

$$= \frac{1}{4}\Gamma\left(\frac{1}{4}\right)$$

To get the last step, the result of Example 11.2, $\Gamma(z) = 2\int_0^\infty x^{2z-1} e^{-x^2} dx$, was used. Now noting that

$$\Gamma(z) = (z-1)!$$

It follows that

$$\Gamma\left(\frac{1}{4}\right)=\left(\frac{1}{4}-1\right)!=\left(-\frac{3}{4}\right)!$$

But we know

$$(z-1)!=\frac{z!}{z} \qquad \text{or} \qquad z!=\frac{(z+1)!}{z+1}$$

This follows from the definition of factorial where $n!=n(n-1)\cdots 1$. Setting $z=-3/4$ we find that

$$\left(-\frac{3}{4}\right)!=\frac{\left(-\frac{3}{4}+1\right)!}{-\frac{3}{4}+1}=4\left(\frac{1}{4}!\right)$$

Therefore

$$\begin{aligned}\int_0^\infty e^{-x^4}dx &= \frac{1}{4}\Gamma\left(\frac{1}{4}\right)\\ &= \frac{1}{4}\left(-\frac{3}{4}\right)!\\ &= \frac{1}{4}4\left(\frac{1}{4}!\right)\\ &= \frac{1}{4}!\end{aligned}$$

EXAMPLE 11.6

Show that $\Gamma[(1/2)-n]\Gamma[(1/2)+n]=(-1)^{-n}\pi$.

SOLUTION

Recalling that

$$\Gamma(z)\Gamma(z-1)=\frac{\pi}{\sin \pi z}$$

Setting $z=(1/2)-n$ we obtain

$$\Gamma\left(\frac{1}{2}-n\right)\Gamma\left(\frac{1}{2}+n\right)=\frac{\pi}{\sin\pi\left(\frac{1}{2}-n\right)}=\frac{\pi}{\sin\left(\frac{\pi}{2}-n\pi\right)}$$

Let's take a look at the denominator for a few values of n:

$$n=0:\sin\left(\frac{\pi}{2}-n\pi\right)=\sin\left(\frac{\pi}{2}\right)=+1$$

$$n=1:\sin\left(\frac{\pi}{2}-n\pi\right)=\sin\left(\frac{\pi}{2}-\pi\right)=\sin\left(-\frac{\pi}{2}\right)=-1$$

$$n=2:\sin\left(\frac{\pi}{2}-n\pi\right)=\sin\left(\frac{\pi}{2}-2\pi\right)=-\sin\left(\frac{3\pi}{2}\right)=+1$$

$$n=3:\sin\left(\frac{\pi}{2}-n\pi\right)=\sin\left(\frac{\pi}{2}-3\pi\right)=-\sin\left(\frac{5\pi}{2}\right)=-1$$

We conclude that $\sin(\pi/2-n\pi)=(-1)^n$. So, we obtain

$$\Gamma\left(\frac{1}{2}-n\right)\Gamma\left(\frac{1}{2}+n\right)=\frac{\pi}{\sin\left(\frac{\pi}{2}-n\pi\right)}=(-1)^{-n}\pi$$

EXAMPLE 11.7
Find $\Gamma(-1/2)$.

SOLUTION
To determine the value of $\Gamma(-1/2)$, we must use the recursion relation together with analytic continuation to extend the definition into the left half of the complex plane [since $\text{Re}(z)<0$]. We already know that

$$\Gamma\left(\frac{1}{2}\right)=\sqrt{\pi}$$

Now we use

$$\Gamma(z+1)=z\Gamma(z)$$

From which it follows that

$$\Gamma\left(-\frac{1}{2}\right)=-\frac{1}{2}\Gamma\left(\frac{1}{2}\right)=-2\sqrt{\pi}$$

More Properties of the Gamma Function

In this section, we list a few properties of the gamma function that can be useful for calculation. A variation of the recursion formula shows that the gamma function is an analytic function except for simple poles which are found in the left-hand plane (see Fig. 7.2 or Figs. 11.2 and 11.3 for an illustration). The following relationship holds:

$$\Gamma(z)=\frac{\Gamma(z+n+1)}{z(z+1)(z+2)\cdots(z+n)} \tag{11.8}$$

This tells us that the gamma function is a meromorphic function. It has simple poles located at $0,-1,-2,-3,\ldots$ but is analytic everywhere else in the complex plane for $\text{Re}(z)>-(n+1)$. In Example 11.8, we'll show how to arrive at formula in Eq. (11.8). When you read the example, note how n is arbitrary, so we can expand $\Gamma(z)$ arbitrarily throughout the complex plane and it will only contain simple poles.

Euler's constant is defined to be

$$\gamma=\lim_{p\to\infty}\left\{1+\frac{1}{2}+\frac{1}{3}+\cdots+\frac{1}{p}-\ln p\right\}=0.5772157\cdots \tag{11.9}$$

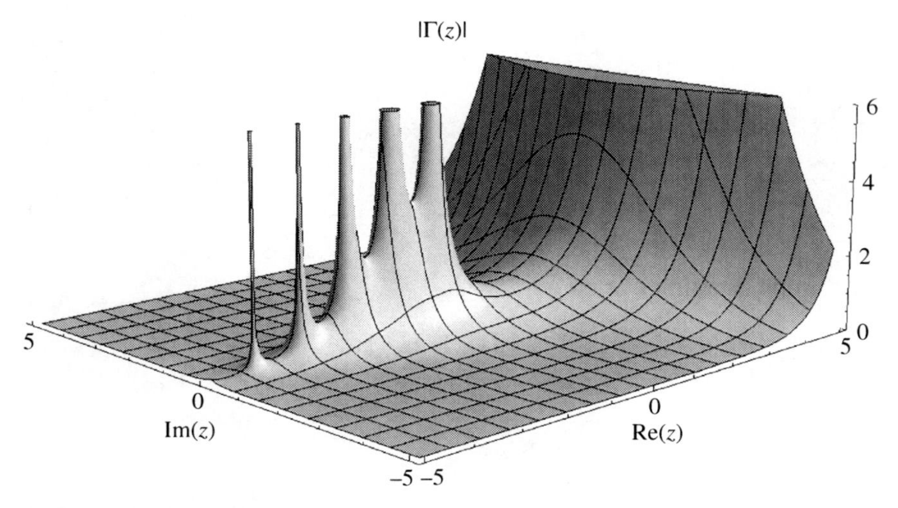

Figure 11.2 An illustration of the simple poles of the gamma function on the negative real axis.

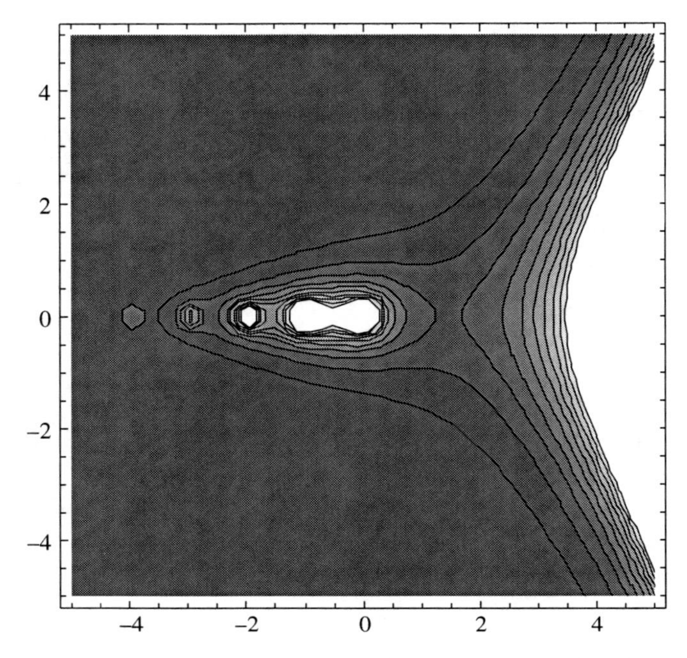

Figure 11.3 A contour plot of the modulus of the gamma function, in the *x*-*y* plane.

Using Euler's constant, we can write down an infinite product representation of the gamma function:

$$\frac{1}{\Gamma(z)} = ze^{\gamma z}\prod_{k=1}^{\infty}\left(1+\frac{z}{k}\right)e^{-z/k} \tag{11.10}$$

Using *Gauss' Π function* $\Pi(z,k)$:

$$\Pi(z,k) = \frac{1\cdot 2\cdot 3\cdots k}{(z+1)(z+2)\cdots(z+k)}k^z \tag{11.11}$$

The gamma function can be written as

$$\Gamma(z+1) = \lim_{k\to\infty}\Pi(z,k) \tag{11.12}$$

The *duplication formula* tells us that

$$2^{2z-1}\Gamma(z)\Gamma\left(z+\frac{1}{2}\right) = \sqrt{\pi}\,\Gamma(2z) \tag{11.13}$$

EXAMPLE 11.8
Show that

$$\Gamma(z) = \frac{\Gamma(z+n+1)}{z(z+1)(z+2)\cdots(z+n)}$$

SOLUTION
Using the recursion formula:

$$\Gamma(z+1) = z\Gamma(z)$$

Therefore it follows that

$$\Gamma(z) = \frac{\Gamma(z+1)}{z}$$

We can apply the recursion formula again, letting $z \to z+1$, which gives

$$\Gamma(z+2) = (z+1)\Gamma(z+1)$$
$$\Rightarrow \Gamma(z+1) = \frac{\Gamma(z+2)}{(z+1)}$$

And so

$$\Gamma(z) = \frac{\Gamma(z+1)}{z} = \frac{\Gamma(z+2)}{z(z+1)}$$

If you carry out this procedure n times, the result follows.

EXAMPLE 11.9
Show that the gamma function is analytic on the right half plane.

SOLUTION
Consider

$$\Gamma_\varepsilon(z) = \int_\varepsilon^{1/\varepsilon} t^{z-1} e^{-t} dt$$

We can immediately show this function is analytic by computing the derivative with respect to $\bar{z}$:

$$\frac{d\Gamma_\varepsilon}{d\bar{z}} = \frac{d}{d\bar{z}} \int_\varepsilon^{1/\varepsilon} t^{z-1} e^{-t} dt = \int_\varepsilon^{1/\varepsilon} \frac{d}{d\bar{z}} (t^{z-1}) e^{-t} dt = 0$$

Taking the limit $\varepsilon \to 0$, the result follows.

EXAMPLE 11.10

Find the residue of the gamma function at the singularity located at $z = -n$, where n is an integer.

SOLUTION

Computing the residue of the function as written in Eq. (11.8), we have

$$\begin{aligned}
\lim_{z \to -n} (z+n)\Gamma(z) &= \lim_{z \to -n} (z+n) \frac{\Gamma(z+n+1)}{z(z+1)(z+2)\cdots(z+n)} \\
&= \lim_{z \to -n} \frac{\Gamma(z+n+1)}{z(z+1)(z+2)\cdots(z+n-1)} \\
&= \frac{\Gamma(1)}{-n(-n+1)(-n+2)\cdots(-1)} \\
&= \frac{\Gamma(1)}{(-1)n(-1)(n-1)(-1)(n-2)\cdots(-1)} \\
&= (-1)^n \frac{\Gamma(1)}{n(n-1)(n-2)\cdots 1} = \frac{(-1)^n}{n!}
\end{aligned}$$

The gamma function is the only meromorphic function which has the following three properties:

- $\Gamma(z+1) = z\Gamma(z)$
- $\Gamma(1) = 1$
- $\log \Gamma(x)$ is convex

EXAMPLE 11.11

Show that the gamma function is logarithmically convex on the real axis.

SOLUTION

Saying that the gamma function is logarithmically convex on the real axis means that if we take it's log and then find the second derivative, then let $z \to x$, the result will be positive.

This is done using the infinite product representation of Eq. (11.10). We reproduce it here for convenience:

$$\frac{1}{\Gamma(z)} = z e^{\gamma z} \prod_{k=1}^{\infty} \left(1 + \frac{z}{k}\right) e^{-z/k}$$

First, we take the logarithm of the left-hand side:

$$\log \frac{1}{\Gamma(z)} = \log \Gamma(z)^{-1} = -\log \Gamma(z)$$

Recalling that the log of a product is the sum of the logs, that is $\log AB = \log A + \log B$, on the right-hand side we find

$$\log \frac{1}{\Gamma(z)} = \log z e^{\gamma z} \prod_{k=1}^{\infty} \left(1 + \frac{z}{k}\right) e^{-z/k}$$

$$= \log z + \gamma z + \sum_{k=1}^{\infty} \left\{ \log\left(1 + \frac{z}{k}\right) - \left(\frac{z}{k}\right) \right\}$$

$$\log \Gamma(z) = -\log z - \gamma z - \sum_{k=1}^{\infty} \log\left(1 + \frac{z}{k}\right) + \sum_{k=1}^{\infty} \left(\frac{z}{k}\right)$$

Therefore

$$\frac{\partial}{\partial z} \log \Gamma(z) = -\frac{1}{z} - \gamma + \left\{ \sum_{k=1}^{\infty} \frac{1}{k} \left(\frac{-1}{1 + \frac{z}{k}} \right) + \left(\frac{1}{k}\right) \right\}$$

$$= -\frac{1}{z} - \gamma + \sum_{k=1}^{\infty} \left\{ \left(\frac{-1}{z + k} \right) + \frac{1}{k} \right\}$$

$$= -\frac{1}{z} - \gamma + \sum_{k=1}^{\infty} \left(\frac{1}{k}\right) \left(\frac{z}{z + k} \right)$$

Computing the second derivative we find

$$\frac{\partial^2}{\partial z^2} \log \Gamma(z) = \frac{1}{z^2} + \sum_{k=1}^{\infty} \frac{1}{(z + k)^2}$$

Evaluating this for real argument:

$$\frac{\partial^2}{\partial x^2} \log \Gamma(x) = \frac{1}{x^2} + \sum_{k=1}^{\infty} \frac{1}{(x + k)^2} > 0 \qquad \text{when } x > 0$$

Showing that the gamma function is logarithmically convex for real argument in the right half plane.

Contour Integral Representation and Stirling's Formula

We close by noting that the gamma function can be written as

$$\frac{1}{\Gamma(z)} = \frac{1}{2\pi i}\oint \frac{e^t}{t^z}\,dt \tag{11.14}$$

The contour used comes in from the negative real axis, goes counterclockwise about the origin and out along the negative real axis, avoiding the branch point at the origin.

The *Stirling approximation* for the gamma function is given by

$$\Gamma(z+1) \approx \sqrt{2\pi}\, e^{-z} z^{z+1/2} \tag{11.15}$$

The Beta Function

The *beta function* is defined by the following integral, where $\mathrm{Re}(m) > 0$ and $\mathrm{Re}(n) > 0$:

$$B(m,n) = \int_0^1 t^{m-1}(1-t)^{n-1}\,dt \tag{11.16}$$

By using the substitution $t = \sin^2\theta$, we can move to polar coordinates and write the beta function in terms of trigonometric functions. First note that

$$dt = 2\sin\theta\cos\theta\, d\theta$$

Using $\cos^2\theta + \sin^2\theta = 1$, we have

$$\begin{aligned}
B(m,n) &= \int_0^1 t^{m-1}(1-t)^{n-1}\,dt \\
&= \int_0^{\pi/2} (\sin^2\theta)^{m-1}(\cos^2\theta)^{n-1}\, 2\sin\theta\cos\theta\, d\theta \\
&= 2\int_0^{\pi/2} (\sin^2\theta)^{2m-1}(\cos^2\theta)^{2n-1}\, d\theta
\end{aligned}$$

The beta function is related to the gamma function via

$$B(m,n) = \frac{\Gamma(m)\Gamma(n)}{\Gamma(m+n)} \tag{11.17}$$

Furthermore, we can write

$$B(p,1-p) = \int_0^\infty \frac{t^{p-1}}{1+t}dt = \Gamma(p)\Gamma(p-1) = \frac{\pi}{\sin p\pi} \tag{11.18}$$

provided that $0 < \mathrm{Re}(p) < 1$.

The Riemann Zeta Function

The Riemann zeta function was studied by Riemann for number theory. It has the following series representation:

$$\zeta(z) = \frac{1}{1^z} + \frac{1}{2^z} + \frac{1}{3^z} + \cdots = \sum_{k=1}^{\infty} \frac{1}{k^z} \tag{11.19}$$

While this series is defined for $\mathrm{Re}(z) > 0$, analytic continuation can be used to extend the zeta function to other values of z. Notice that we can write Eq. (11.19) as

$$\zeta(z) = \sum_{k=1}^{\infty} \frac{1}{k^z} = \sum_{k=1}^{\infty} \frac{1}{e^{\log k^z}} = \sum_{k=1}^{\infty} \frac{1}{e^{z \log k}} = \sum_{k=1}^{\infty} e^{-z \log k}$$

The zeta function can be defined in terms of the gamma function via the following relationship:

$$\zeta(z) = \frac{1}{\Gamma(z)} \int_0^\infty \frac{t^{z-1}}{e^t + 1} dt \tag{11.20}$$

Another way to relate the zeta and gamma functions—and to define a recursion relation for the zeta function—is by using

$$\zeta(1-z) = 2^{1-z}\pi^{-z}\Gamma(z)\cos\left(\frac{\pi z}{2}\right)\zeta(z) \tag{11.21}$$

When $0 < x < 1$ a plot of $|\zeta(z)|$ shows a series of ridges located at different points along the imaginary axis. These ridges are characterized by the fact that they are monotonically decreasing. This is illustrated in Fig. 11.4. A plot over a wider range of the real axis is shown in Fig. 11.5, and a contour plot of the zeta function is shown in Fig. 11.6.

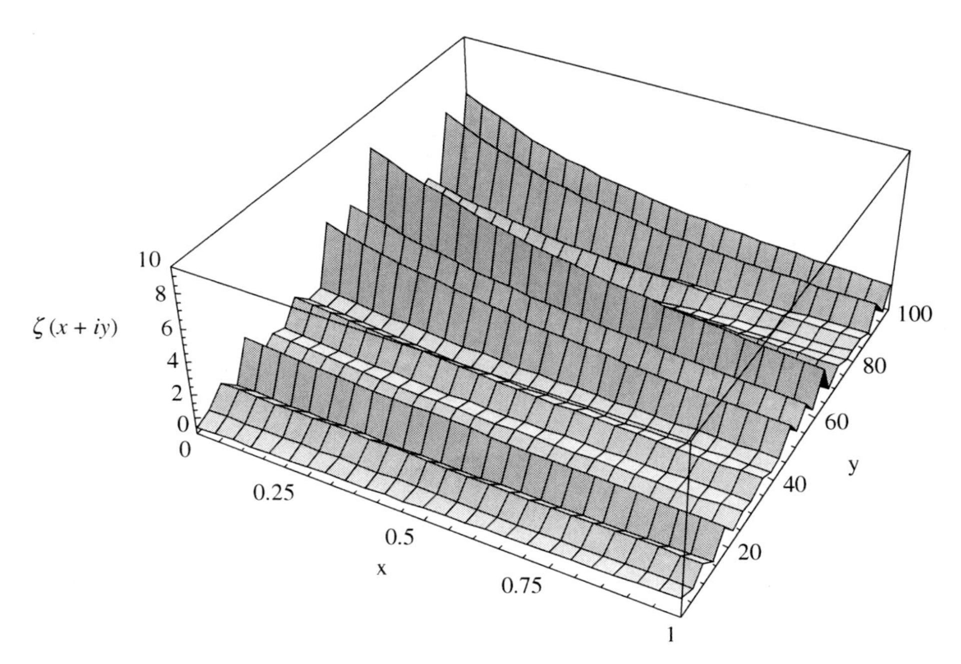

Figure 11.4 A plot of the modulus of the Riemann zeta function. $|\zeta(x+iy)|$ is characterized by monotonically decreasing ridges.

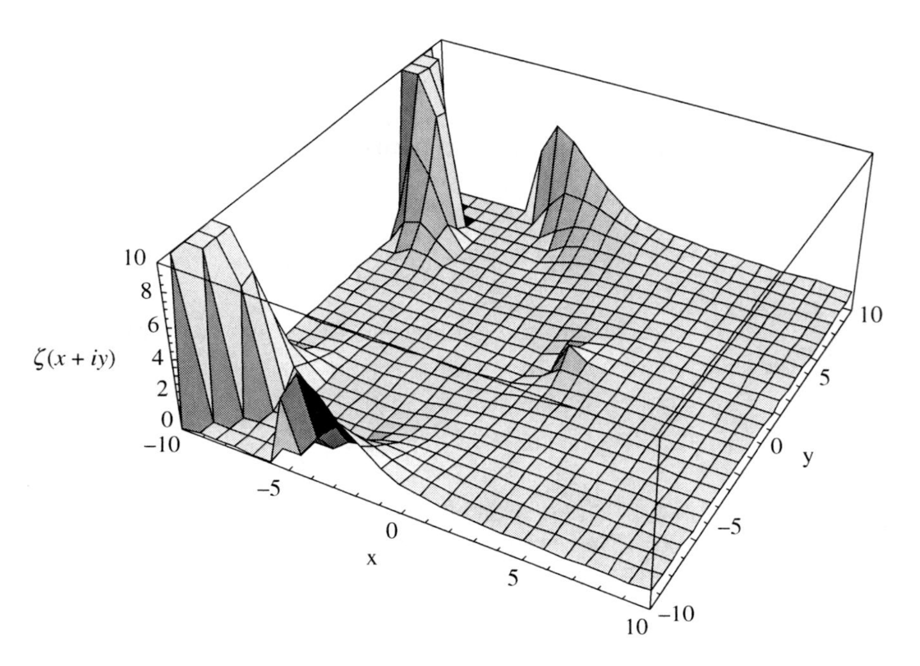

Figure 11.5 A wider view of the modulus of the Riemann zeta function.

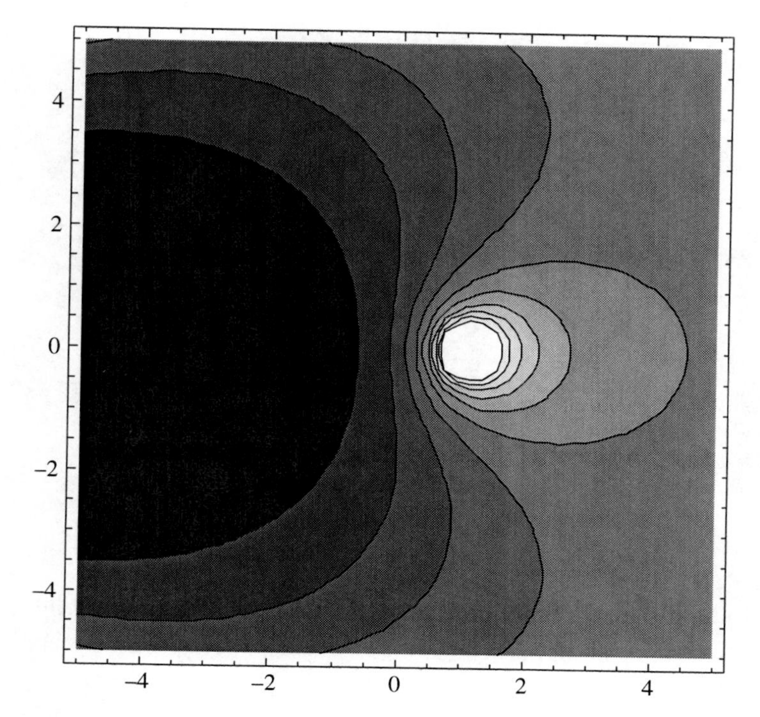

Figure 11.6 A contour plot of the Riemann zeta function in the x-y plane.

An unproven conjecture by Riemann is that all of the zeros of $\zeta(z)$ can be found on the line $\mathrm{Re}(z) = 1/2$. We show a plot of $|\zeta(1/2 + iy)|$ in Fig. 11.7, where you can see the zeros of the function for $0 \le y \le 50$. The real and imaginary parts of $\zeta(1/2 + iy)$ are illustrated in Figs. 11.8 and 11.9, respectively. While Riemann's conjecture has yet to be proven, Hardy demonstrated that there are infinitely many zeros along the line $\mathrm{Re}(z) = 1/2$.

In Fig. 11.10, we consider the Riemann zeta function for real argument. A plot of $\zeta(x)$ shows an asymptote at $x = 1$ where the function blows up.

Like the gamma function, the Riemann zeta function has a representation in terms of infinite products. This is given by

$$\frac{1}{\zeta(z)} = \left(1 - \frac{1}{2^z}\right)\left(1 - \frac{1}{3^z}\right)\left(1 - \frac{1}{5^z}\right)\cdots = \prod_p \left(1 - \frac{1}{p^z}\right) \tag{11.22}$$

The interesting (and maybe somewhat mysterious) feature of Eq. (11.22) is that the product is taken over all *positive primes p*.

EXAMPLE 11.12

Is the Riemann zeta function analytic in a region of the complex plane for which $\mathrm{Re}(z) \ge 1 + \varepsilon$ where $\varepsilon > 0$?

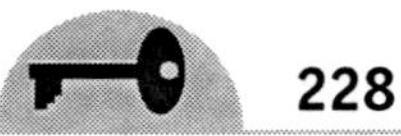

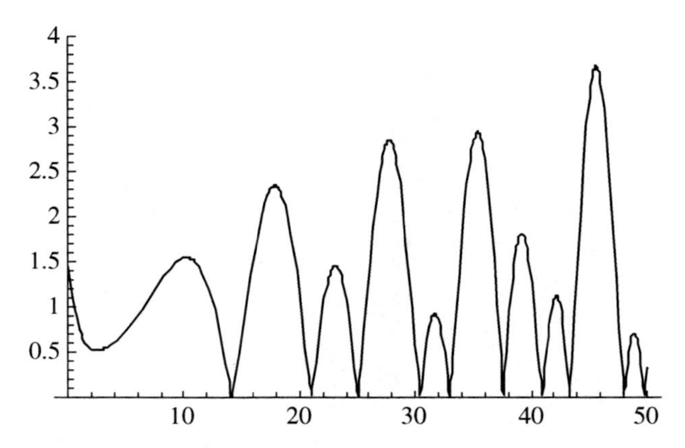

Figure 11.7 A plot of $|\zeta(1/2+iy)|$ along the imaginary axis, showing the zeros of the zeta function.

SOLUTION

To determine if the zeta function is analytic, take the series representation

$$\zeta(z) = \sum_{k=1}^{\infty} \frac{1}{k^z}$$

If $\mathrm{Re}(z) \geq 1+\varepsilon$, then this means that $x \geq 1+\varepsilon$. And so

$$\left|\frac{1}{k^z}\right| = \left|\frac{1}{e^{z\ln k}}\right| = \frac{1}{e^{x\ln k}} = \frac{1}{k^x} \leq \frac{1}{k^{1+\varepsilon}}$$

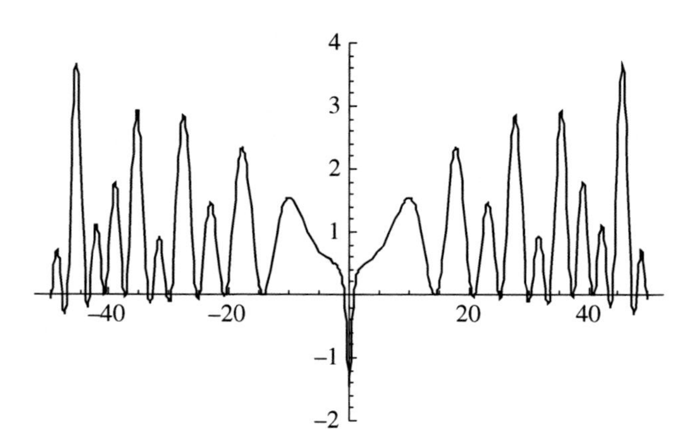

Figure 11.8 A plot of the real part of the zeta function for $x = 1/2$.

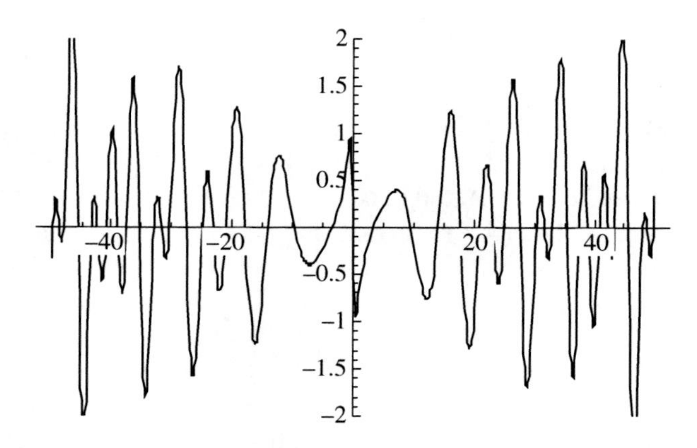

Figure 11.9 A plot of the imaginary part of the zeta function for $x = 1/2$.

The series

$$\sum \frac{1}{k^{1+\varepsilon}}$$

is convergent, so by the Weierstrass M-test $\zeta(z) = \sum_{k=1}^{\infty}(1/k^{z})$ is convergent as well. In fact, the zeta function converges uniformly for $\text{Re}(z) \geq 1 + \varepsilon$. This proves that if $\text{Re}(z) \geq 1 + \varepsilon$, the zeta function is analytic.

The zeta function has a single simple pole located at $z = 1$. The residue corresponding to this singularity is one.

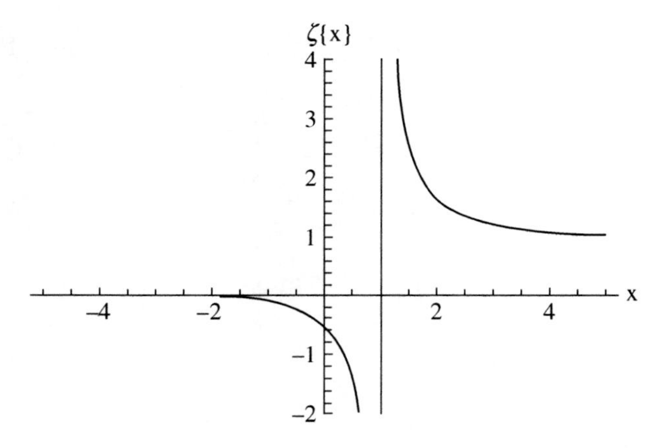

Figure 11.10 A plot of the Riemann zeta function with real argument. Note the asymptote at $x = 1$ where the function blows up.

Summary

In this chapter, we introduced three special functions from complex analysis. These include the gamma function, the beta function, and the Riemann zeta function. These functions can be represented in different ways, using series, infinite products, or integral representations (in the case of the gamma and beta functions).

Quiz

1. Using Eqs. (11.1) and (11.2) calculate 0!.
2. Consider Gauss' Π function. What is $\lim_{k\to\infty}\Pi(3,k)$?
3. Compute $\int_0^4 y^{3/2}\sqrt{16-y^2}\,dy$ using gamma or beta functions.
4. Find $\int_0^4 \sqrt{x}\sqrt{4-x}\,dx$ using gamma and beta functions.
5. Using the gamma function, calculate $\int_0^\infty \cos(t^3)\,dt$.
6. Find an expression for $\Gamma(z)\Gamma(-z)$ and use it to calculate $\Gamma\left(-\frac{1}{2}\right)$ by considering $\Gamma(z)\Gamma(1-z)$.
7. Considering that $f(z)=(1+z)^\alpha=\sum_{n=0}^{\infty}\frac{\alpha!}{n!(\alpha-n)!}$, find an expression for $d^nf/dz^n\big|_{z=0}$ in terms of gamma functions.
8. How can you prove that $\zeta(1)$ is divergent?
9. Evaluate $\frac{\Gamma(z+n)}{n^z\Gamma(n)}$.
10. Describe the points at which $f(z)=\zeta(z)-\frac{1}{z-1}$ is not analytic.

CHAPTER 12

Boundary Value Problems

Complex analysis can be utilized to solve partial differential equations. In this chapter, we consider *Dirichlet problems,* which involve the specification of a function that solves Laplace's equation in a region R of the x-y plane and takes on prescribed values on the boundary of the region which a curve is enclosing R that we denote by C. A *Neumann problem* is one that specifies the derivative of the function on the boundary. Conformal mapping can be used to arrive at a solution to these types of problems.

Laplace's Equation and Harmonic Functions

Let's review the concept of a harmonic function, which was introduced in Chap. 3. We say that a function $\phi(x, y)$ is a harmonic function in a region R of the x-y plane if it satisfies Laplace's equation:

$$\nabla^2 \phi = \frac{\partial^2 \phi}{\partial x^2} + \frac{\partial^2 \phi}{\partial y^2} = 0 \tag{12.1}$$

We often use the shorthand notation $\phi_{xx} = \partial^2 \phi / \partial x^2$ and $\phi_{yy} = \partial^2 \phi / \partial y^2$ to indicate partial derivatives. Recall that if a complex function $f(z) = u(x, y) + iv(x, y)$ is analytic in a region R then it follows that $u(x, y)$ and $v(x, y)$ are harmonic functions. Moreover, $u(x, y)$ and $v(x, y)$ are harmonic conjugates, meaning that one can be determined from the other by integration and the addition of an arbitrary constant (see Chap. 3).

In polar coordinates, Eq. (12.1) becomes

$$\frac{\partial^2 u}{\partial r^2} + \frac{1}{r}\frac{\partial u}{\partial r} + \frac{1}{r^2}\frac{\partial^2 u}{\partial \theta^2} = 0 \tag{12.2}$$

The first two examples review the concept of harmonic conjugate.

EXAMPLE 12.1

Find the harmonic conjugate of $u(x, y) = x/(x^2 + y^2)$.

SOLUTION

First we compute the derivatives of the function with respect to x and y:

$$\frac{\partial u}{\partial x} = \frac{\partial}{\partial x}\left(\frac{x}{x^2 + y^2}\right)$$

This derivative can be computed using the rule for quotients:

$$\left(\frac{f}{g}\right)' = \frac{f'g - g'f}{g^2}$$

Setting $f = x$ and $g = x^2 + y^2$ we have

$$f' = 1 \quad \text{and} \quad g' = 2x$$

Therefore

$$\frac{\partial u}{\partial x} = \frac{\partial}{\partial x}\left(\frac{x}{x^2 + y^2}\right) = \frac{x^2 + y^2 - x(2x)}{(x^2 + y^2)^2} = \frac{y^2 - x^2}{(x^2 + y^2)^2}$$

A similar procedure yields the derivative with respect to y:

$$\frac{\partial u}{\partial y} = \frac{\partial}{\partial y}\left(\frac{x}{x^2 + y^2}\right) = \frac{-2xy}{(x^2 + y^2)^2}$$

The Cauchy-Riemann equations tell us that

$$\frac{\partial u}{\partial x} = \frac{\partial v}{\partial y} \qquad \frac{\partial v}{\partial x} = -\frac{\partial u}{\partial y}$$

Hence we have the following relationship:

$$\frac{\partial v}{\partial x} = \frac{\partial u}{\partial y} = \frac{2xy}{(x^2+y^2)^2}$$

$$\Rightarrow v(x,y) = 2y\int \frac{x}{(x^2+y^2)^2}dx + F(y)$$

If we let $s = x^2 + y^2 \Rightarrow ds = 2xdx$ then

$$v(x,y) = y\int \frac{ds}{s^2} + F(y) = -y\frac{1}{s} + F(y) = \frac{-y}{x^2+y^2} + F(y)$$

Now, we also have the other Cauchy-Riemann equation at our disposal, which says

$$\frac{\partial v}{\partial y} = \frac{\partial u}{\partial x} = \frac{y^2 - x^2}{(x^2+y^2)^2} \tag{12.3}$$

If we take the derivative with respect to y of $v(x,y) = [-y/(x^2+y^2)^2] + F(y)$, we obtain

$$\frac{\partial v}{\partial y} = \frac{\partial}{\partial y}\left(\frac{-y}{(x^2+y^2)}\right) + F'(y) = \frac{y^2-x^2}{(x^2+y^2)^2} + F'(y)$$

Comparison with Eq. (12.3) tells us that $F'(y) = 0$, which means that $F(y)$ is some constant. Choosing it to be 0, we find that the harmonic conjugate to $u(x, y)$ is

$$v(x,y) = \frac{-y}{(x^2+y^2)^2}$$

EXAMPLE 12.2

Find the harmonic conjugate of $u(r,\theta) = \ln r$.

SOLUTION

To solve this problem, we recall the form of the Cauchy-Riemann equations in polar coordinates, given in Eq. (3.27):

$$\frac{\partial u}{\partial r} = \frac{1}{r}\frac{\partial v}{\partial \theta} \qquad \frac{\partial v}{\partial r} = -\frac{1}{r}\frac{\partial u}{\partial \theta}$$

Since u is a function of r only, we can immediately deduce that

$$\frac{\partial u}{\partial \theta} = 0 \Rightarrow v(r,\theta) = A + f(\theta)$$

where A is some constant. Now

$$\frac{\partial u}{\partial r} = \frac{\partial}{\partial r} \ln r = \frac{1}{r}$$

But by the Cauchy-Riemann equations, we have that $\partial u/\partial r = (1/r)(\partial v/\partial \theta)$. Therefore, $\partial v/\partial \theta = 1$ which we can use to write

$$v(r,\theta) = \int d\theta = \theta + B$$

where B is some constant. Comparison with $v(r,\theta) = A + f(\theta)$ leads us to

$$f(\theta) = \theta$$

Ignoring the constants of integration, we conclude that the harmonic conjugate of $u(r,\theta) = \ln r$ is

$$v(r,\theta) = \theta$$

Solving Boundary Value Problems Using Conformal Mapping

In this section, we apply conformal mapping techniques to the solution of boundary value problems with Dirichlet and Neumann boundary conditions. First, we state *Poisson's formulas*, which give the solutions to the Dirichlet problem on the unit disk and for the upper half plane.

1. Let C be the unit circle and R its interior. Suppose that $f(r,\theta)$ is harmonic in R and that it assumes the value $g(\theta)$ on the curve C, that is, $f(1,\theta) = g(\theta)$. Then the solution to Laplace's equation on the unit disk is given by *Poisson's formula for a circle* which states that

$$f(r,\theta) = \frac{1}{2\pi} \int_0^{2\pi} \frac{(1-r^2)g(\phi)\,d\phi}{1 - 2r\cos(\theta - \phi) + r^2} \tag{12.4}$$

Next we consider a function $f(x, y)$, which is harmonic in the upper half plane $y > 0$ and assumes the value $f(x, y) = g(x)$ on the boundary, which in this case is the x axis, that is, $-\infty < x < \infty$. The solution to Laplace's equation for the upper half plane is given by

$$f(x, y) = \frac{1}{\pi}\int_{-\infty}^{\infty} \frac{y\, g(s)}{y^2 + (x-s)^2}\, ds \tag{12.5}$$

We call Eq. (12.5) *Poisson's formula for the half plane.*

We can solve a wide variety of boundary value problems for a simply connected region R by using conformal mapping techniques. The idea is to map the region R to the unit disk or to the half plane. The mapping function that we use must be analytic. There are three steps involved in obtaining a solution:

- Use a conformal transformation to map the boundary value problem for a region R to a boundary value problem on the unit disk or half plane.
- Solve the problem using Eq. (12.4) or (12.5).
- Find the inverse of the solution (that is apply the inverse of the conformal mapping) to write down the solution in the region R.

Remember that a simply connected region is one that includes no singularities. While the mapping of the region R to a region R' in the w plane must be conformal, the mapping of the boundary does not have to be conformal.

Three theorems are useful for solving these types of problems. The first tells us that a map $w = f(z)$ has a unique inverse. The second tells us that a harmonic function is transformed into a harmonic function under an analytic mapping $w = f(z)$, and finally we learn that if the boundary value of a function is constant in the z plane, then it is in the w plane as well.

THEOREM 12.1

Let $w = f(z)$ be analytic in a region R of the z plane. Then there exists a unique inverse $z = g(w)$ in R if $f'(z) \neq 0$ in R.

THEOREM 12.2

Let $U(x, y)$ be a function in some region R of the z plane and suppose that $w = f(z)$ is analytic with $f'(z) \neq 0$. Then if $\phi(u, v)$ is harmonic in the w plane where $U(z) = \phi(w)$ are related by $w = f(z)$, then $U(x, y)$ is harmonic.

THEOREM 12.3

Let $U(x, y) = A$, where A is a constant on the boundary or part of the boundary C of a region R in the z plane. Then its image $\phi(u, v) = A$ on the boundary of the region R' of the w plane. If the normal derivative with respect to the boundary $\partial U / \partial n = 0$ in the z plane, then $\partial\phi / \partial n = 0$ on the boundary of R' in the w plane as well.

We prove theorem 12.2. We know that $\phi(u,v)$ is harmonic in the w plane. This tells us that

$$\phi_{uu} + \phi_{vv} = 0$$

The coordinates u and v are functions of x and y. So we can take the derivatives of $U(x,y)$ using the chain rule and the fact that $U(z) = \phi(w)$:

$$U_x = \phi_u u_x + \phi_v v_x \qquad U_y = \phi_u u_y + \phi_v v_y$$
$$\Rightarrow U_{xx} = \phi_{uu} u_x^2 + 2\phi_{uv} u_x v_x + \phi_{vv} v_x^2 + \phi_u u_{xx} + \phi_v v_{xx}$$
$$U_{yy} = \phi_{uu} u_y^2 + 2\phi_{uv} u_y v_y + \phi_{vv} v_y^2 + \phi_u u_{yy} + \phi_v v_{yy}$$

But the Cauchy-Riemann equations are also satisfied by u and v since we assumed that $w = f(z)$ is analytic. So

$$u_x = v_y \qquad u_y = -v_x$$

Moreover, since the transformation is analytic, and $f = u + iv$, the coordinate functions u and v are harmonic, that is, $u_{xx} + u_{yy} = v_{xx} + v_{yy} = 0$. It follows that

$$U_{xx} + U_{yy} = (\phi_{uu} + \phi_{vv})u_x^2 + (\phi_{uu} + \phi_{vv})u_y^2 + \phi_u(u_{xx} + u_{yy}) + \phi_v(v_{xx} + v_{yy}) = 0$$

This concludes the proof, which showed that if $\phi(u,v)$ is harmonic in the w plane, then $U(x,y)$ is harmonic in the z plane.

Summarizing, the three theorems stated above tell us that given a mapping $w = f(z)$ that is analytic and that takes a region R of the z plane to a region R' of the w plane, we have an inverse mapping $z = w^{-1}(z)$. If $\phi(u,v)$ solves $\phi_{uu} + \phi_{vv} = 0$ in some region R' of the w plane and $\phi(w) = U(z)$ then it follows that $U(x,y)$ satisfies $U_{xx} + U_{yy} = 0$.

EXAMPLE 12.3

Consider the quarter plane defined by $0 < x < \infty, 0 < y < \infty$. Solve Laplace's equation:

$$\frac{\partial^2 f}{\partial x^2} + \frac{\partial^2 f}{\partial y^2} = 0$$

with Dirichlet boundary conditions given by $f(0,y) = 1, f(x,0) = 0$.

SOLUTION

We can apply conformal mapping to this problem by recalling that the map $w = z^n$ increases angles by n meaning that $\theta \to n\theta$. For practice, let's look at the quarter plane, shown in Fig. 12.1.

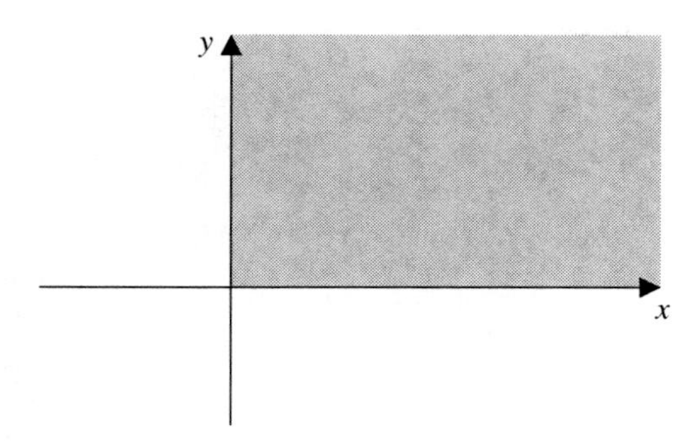

Figure 12.1 The problem in Example 12.3 is specified in the quarter plane $0 < x < \infty, 0 < y < \infty$.

We want a mapping that will take the angle $\pi/2 \to \pi$, so that we can double the region and use Poisson's formula for the half plane. Recalling our discussion in Chap. 9, the transformation or mapping that will accomplish this is:

$$w = z^2$$

which is given by

$$w = (x + iy)^2 = x^2 - y^2 + i2xy$$
$$\Rightarrow u = x^2 - y^2 \qquad v = 2xy$$

Notice that as $x \to \infty$ for fixed $y \neq 0$ that $v \to +\infty$. It also the case that as $y \to \infty$ for fixed $x \neq 0$ that $v \to +\infty$. Since $0 < x < \infty$, $0 < y < \infty$, the other extreme is $x = 0$ or $y = 0$ and in either case $v = 0$. So, we see that we have $0 < v < \infty$. On the other hand, suppose that $x = 0, y \to \infty$. Then $u \to -\infty$. If $y = 0$, $x \to \infty$ then $u \to +\infty$. So we have $-\infty < u < \infty$.

So this map successfully generates the entire upper half plane from the quarter plane, as also explained in Chap. 9. The w plane is illustrated in Fig. 12.2.

To solve the problem in the upper half plane, we use Eq. (12.5). In the w plane this is

$$\phi(u, v) = \frac{1}{\pi} \int_{-\infty}^{\infty} \frac{v\, g(s)}{v^2 + (u - s)^2} ds$$

The boundary condition given is $f(0, y) = 1, f(x, 0) = 0$ (in the z plane). Notice this is a Dirichlet boundary condition since the value of the function is being specified on the boundary. These boundary conditions translate into $g(u,0) = 1$ for $-\infty < u < 0$ and $g(u,0) = 0$ for $0 < u < \infty$. This is because when $x = 0$, we have $u = -y^2, v = 0$. Given the range of $0 < y < \infty$ this fixes the boundary condition to 1 when $-\infty < u < 0$ and 0 when $0 < u < \infty$.

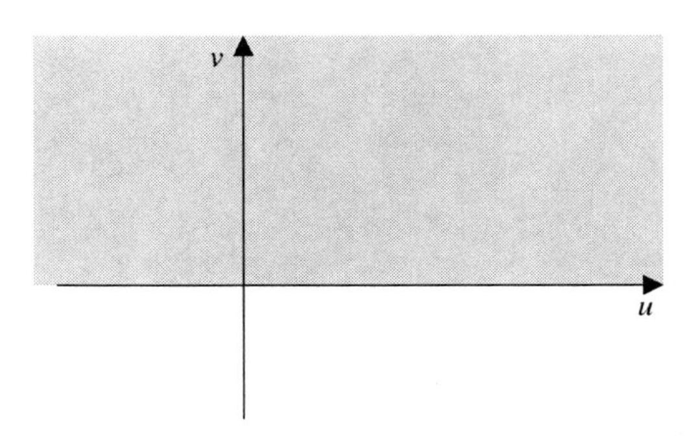

Figure 12.2 The mapping $w = z^2$ transforms the quarter plane shown in Fig. 12.1 into the entire upper half plane.

Hence

$$\phi(u,v) = \frac{1}{\pi}\int_{-\infty}^{\infty}\frac{v\,g(s)}{v^2+(u-s)^2}ds = \frac{1}{\pi}\int_{-\infty}^{0}\frac{v\,g(s)}{v^2+(u-s)^2}ds + \frac{1}{\pi}\int_{0}^{\infty}\frac{v\,g(s)}{v^2+(u-s)^2}ds$$

$$= \frac{1}{\pi}\int_{-\infty}^{0}\frac{v}{v^2+(u-s)^2}ds = \frac{1}{\pi}\tan^{-1}\left(\frac{s-u}{v}\right)\Bigg|_{-\infty}^{0} = \frac{1}{\pi}\tan^{-1}\left(\frac{-u}{v}\right) - \frac{1}{\pi}\left(-\frac{\pi}{2}\right)$$

$$= \frac{1}{2} - \frac{1}{\pi}\tan^{-1}\left(\frac{u}{v}\right)$$

Now

$$\lim_{v\to 0, u=u_0>0}\frac{1}{\pi}\tan^{-1}\left(\frac{u}{v}\right) = \frac{1}{2}$$

So $\phi(u,0) = 0$ when $u > 0$ as required. Secondly:

$$\lim_{v\to 0, u=u_0<0}\frac{1}{\pi}\tan^{-1}\left(\frac{u}{v}\right) = -\frac{1}{2}$$

which leads to $\phi(u,0) = 1$ when $u < 0$. Inverting the transformation gives

$$\phi(u,v) = \frac{1}{2} - \frac{1}{\pi}\tan^{-1}\left(\frac{u}{v}\right)$$

$$\Rightarrow f(x,y) = \frac{1}{2} - \frac{1}{\pi}\tan^{-1}\left(\frac{x^2-y^2}{2xy}\right)$$

We have

$$f(0,y) = \lim_{x\to 0, y>0} f(x,y) = \lim_{x\to 0, y>0} \left\{ \frac{1}{2} - \frac{1}{\pi} \tan^{-1}\left(\frac{x^2 - y^2}{2xy} \right) \right\}$$

$$= \frac{1}{2} - \frac{1}{\pi} \lim_{x\to 0, y>0} \tan^{-1}\left(\frac{x^2 - y^2}{2xy} \right) = \frac{1}{2} - \frac{1}{\pi}\left(-\frac{\pi}{2} \right) = 1$$

and

$$f(x,0) = \lim_{y\to 0, x>0} f(x,y) = \lim_{y\to 0, x>0} \left\{ \frac{1}{2} - \frac{1}{\pi} \tan^{-1}\left(\frac{x^2 - y^2}{2xy} \right) \right\}$$

$$= \frac{1}{2} - \frac{1}{\pi} \lim_{y\to 0, x>0} \tan^{-1}\left(\frac{x^2 - y^2}{2xy} \right) = \frac{1}{2} - \frac{1}{\pi}\left(\frac{\pi}{2} \right) = 0$$

Therefore the boundary conditions in the problem are satisfied.

EXAMPLE 12.4

Consider the unit disk with boundary values specified by

$$g(\theta) = \begin{cases} 1 & 0 < \theta < \pi \\ 0 & \pi < \theta < 2\pi \end{cases}$$

and find a solution to Laplace's equation inside the unit disk.

SOLUTION

This can be done directly using Poisson's formula. Denoting the solution by $f(r,\theta)$ we have

$$f(r,\theta) = \frac{1}{2\pi} \int_0^{2\pi} \frac{g(\phi)}{1 - 2r\cos(\theta - \phi) + r^2} d\phi$$

$$= \frac{1}{2\pi} \int_0^{\pi} \frac{1}{1 - 2r\cos(\theta - \phi) + r^2} d\phi = 1 - \frac{1}{\pi} \tan^{-1}\left(\frac{2r\sin\theta}{1 - r^2} \right)$$

Alternatively, if you would prefer to avoid the integral, the problem can be solved by mapping the unit disk to the upper half plane, as shown in Fig. 12.3. The points *A, B, C, D,* and *E* map to the points A', B', C', D', E', respectively.

The following transformation will work:

$$w = i\left(\frac{1 - z}{1 + z} \right)$$

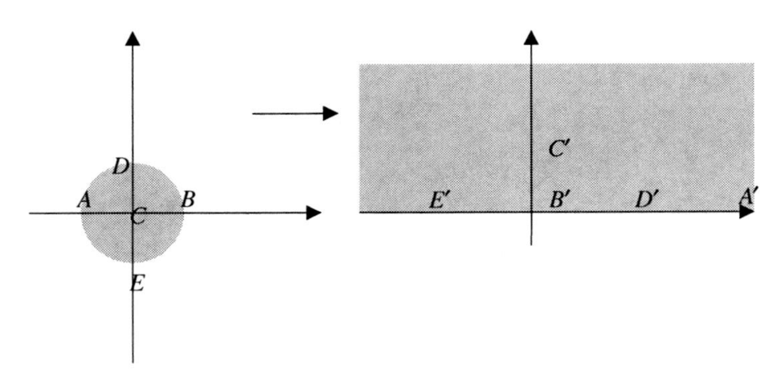

Figure 12.3 In Example 12.4 we map the unit disk to the upper half plane.

Here is how it maps the points A, B, C, D, and E in the figure:

$$A = -1 \qquad \Rightarrow A' = i\left(\frac{1+1}{1-1}\right) \to \infty$$

$$B = 1 \qquad \Rightarrow B' = i\left(\frac{1-1}{1+1}\right) = 0$$

$$C = 0 \qquad \Rightarrow C' = i\left(\frac{1-0}{1+0}\right) = i$$

$$D = i \qquad \Rightarrow D' = i\left(\frac{1-i}{1+i}\right) = +1$$

The solution to the Dirichlet problem in the upper half plane is given by

$$\Phi = 1 - \frac{1}{\pi}\tan^{-1}\left(\frac{u}{v}\right)$$

Now notice that

$$\begin{aligned} w &= i\left(\frac{1-z}{1+z}\right) = i\left(\frac{1-x-iy}{1+x+iy}\right) \\ &= i\left(\frac{1-x-iy}{1+x+iy}\right)\left(\frac{1+x-iy}{1+x-iy}\right) \\ &= i\left(\frac{1-x^2-i2y-y^2}{(1+x)^2+y^2}\right) = \frac{2y}{(1+x)^2+y^2} + i\frac{1-(x^2+y^2)}{(1+x)^2+y^2} \end{aligned}$$

So we have

$$u = \frac{2y}{(1+x)^2 + y^2} \qquad \text{and} \qquad v = \frac{1-(x^2+y^2)}{(1+x)^2+y^2}$$

These functions can be rewritten in terms of polar coordinates using $x = r\cos\theta$, $y = r\sin\theta$:

$$u = \frac{2r\sin\theta}{(1+r\cos\theta)^2 + r^2\sin^2\theta} \qquad \text{and} \qquad v = \frac{1-r^2}{(1+r\cos\theta)^2 + r^2\sin^2\theta},$$

$$\Rightarrow \frac{u}{v} = \frac{2r\sin\theta}{1-r^2}$$

Hence

$$f = 1 - \frac{1}{\pi}\tan^{-1}\left(\frac{2r\sin\theta}{1-r^2}\right)$$

EXAMPLE 12.5

Consider a disk $0 \le r < R$ and solve Laplace's equation with Neumann boundary conditions given by

$$\frac{\partial u}{\partial r}(R,\theta) = g(\theta)$$

$$u(r,\theta) \text{ is bounded as } r \to 0$$

and

$$\int_0^{2\pi} g(\theta)\,d\theta = \int_0^{2\pi}\int_0^R u(r,\theta)\,r\,dr\,d\theta = 0$$

SOLUTION

Laplace's equation in polar coordinates is given by

$$\frac{1}{r}\frac{\partial}{\partial r}\left(r\frac{\partial u}{\partial r}\right) + \frac{1}{r^2}\frac{\partial^2 u}{\partial r^2} = 0 \qquad 0 \le r < a \qquad 0 \le \theta \le 2\pi$$

$$u(a,\theta) = g(\theta), u(r,\theta) \text{ bounded as } r \to 0$$

We solved this problem for $0 \le r < 1$ and different boundary conditions using separation of variables in Example 7.1. We can use the same general solution found there and apply the present boundary conditions. We had found that

$$u_n(r,\theta) = (c_n r^n + c_{-n} r^{-n})(a_n \cos n\theta + b_n \sin n\theta)$$

The requirement that $u(r,\theta)$ remain bounded as $r \to 0$ means that

$$c_{-n} = 0$$

Hence we take

$$u_n(r,\theta) = r^n (a_n \cos n\theta + b_n \sin n\theta)$$

The total solution is a superposition of all the solutions $u_n(r,\theta)$:

$$u(r,\theta) = \sum_{n=0}^{\infty} u_n(r,\theta) = \sum_{n=0}^{\infty} r^n (a_n \cos n\theta + b_n \sin n\theta) = a_0 + \sum_{n=1}^{\infty} r^n (a_n \cos n\theta + b_n \sin n\theta)$$

The derivative of this expression with respect to the radial coordinate is

$$\frac{\partial u}{\partial r}(r,\theta) = \sum_{n=1}^{\infty} n\, r^{n-1} (a_n \cos n\theta + b_n \sin n\theta)$$

So we have

$$g(\theta) = \frac{\partial u}{\partial r}(R,\theta) = \sum_{n=1}^{\infty} n\, R^{n-1} (a_n \cos n\theta + b_n \sin n\theta)$$

This satisfies

$$\begin{aligned}
\int_0^{2\pi} g(\theta)\, d\theta &= \int_0^{2\pi} \left\{ \sum_{n=1}^{\infty} n\, R^{n-1} (a_n \cos n\theta + b_n \sin n\theta) \right\} d\theta \\
&= \sum_{n=1}^{\infty} n\, R^{n-1} \left(a_n \int_0^{2\pi} \cos n\theta\, d\theta + b_n \int_0^{2\pi} \sin n\theta\, d\theta \right) = 0
\end{aligned}$$

as required. Multiplying through $g(\theta)$ by $\sin m\theta$ and integrating we obtain

$$\begin{aligned}
\int_0^{2\pi} g(\theta) \sin m\theta\, d\theta &= \int_0^{2\pi} \left\{ \sum_{n=1}^{\infty} n\, R^{n-1} (a_n \cos n\theta \sin m\theta + b_n \sin n\theta \sin m\theta) \right\} d\theta \\
&= \sum_{n=1}^{\infty} n\, R^{n-1} \left(a_n \int_0^{2\pi} \cos n\theta \sin m\theta\, d\theta + b_n \int_0^{2\pi} \sin n\theta \sin m\theta\, d\theta \right) \\
&= \sum_{n=1}^{\infty} (nR^{n-1})(b_n \pi \delta_{mn}) \\
&= mR^{m-1} \pi\, b_m
\end{aligned}$$

(see Example 7.1). Therefore the coefficient in the expansion is given by

$$b_n = \frac{1}{nR^{n-1}\pi}\int_0^{2\pi} g(\theta)\sin n\theta\, d\theta$$

Now we repeat the process, multiplying through $g(\theta)$ by $\cos m\theta$ and integrating

$$\int_0^{2\pi} g(\theta)\cos m\theta\, d\theta = \int_0^{2\pi}\left\{\sum_{n=1}^{\infty} n\,R^{n-1}(a_n \cos n\theta \cos m\theta + b_n \sin n\theta \cos m\theta)\right\}d\theta$$

$$= \sum_{n=1}^{\infty} n\,R^{n-1}\left(a_n \int_0^{2\pi} \cos n\theta \cos m\theta\, d\theta + b_n \int_0^{2\pi} \sin n\theta \cos m\theta\, d\theta\right)$$

$$= \sum_{n=1}^{\infty} (nR^{n-1})(a_n \pi \delta_{mn})$$

$$= mR^{m-1}\pi\, a_m$$

Hence

$$a_n = \frac{1}{nR^{n-1}\pi}\int_0^{2\pi} g(\theta)\cos n\theta\, d\theta$$

The problem also requires that the solution satisfy

$$\int_0^{2\pi}\int_0^{R} u(r,\theta)\, r\, dr\, d\theta = 0$$

We have

$$\int_0^{2\pi}\int_0^{R} u(r,\theta)\, r\, dr\, d\theta = \int_0^{2\pi}\int_0^{R}\left\{a_0 + \sum_{n=1}^{\infty} r^n (a_n \cos n\theta + b_n \sin n\theta)\right\} r\, dr\, d\theta$$

$$= \int_0^{R} r\, dr\left\{\int_0^{2\pi} a_0\, d\theta + \sum_{n=1}^{\infty} r^n\left(a_n \int_0^{2\pi} \cos n\theta\, d\theta + b_n \int_0^{2\pi} \sin n\theta\, d\theta\right)\right\}$$

$$= a_0 \int_0^{R} r\, dr \int_0^{2\pi} d\theta = a_0 R(2\pi)$$

Therefore $a_0 = 0$ and we can take

$$u(r,\theta) = \sum_{n=1}^{\infty} r^n (a_n \cos n\theta + b_n \sin n\theta)$$

Using $a_n = [1/(nR^{n-1}\pi)]\int_0^{2\pi} g(\theta)\cos n\theta\, d\theta$ and $b_n = [1/(nR^{n-1}\pi)]\int_0^{2\pi} g(\theta)\sin n\theta\, d\theta$, the solution can be written as

$$\begin{aligned}
u(r,\theta) &= \sum_{n=1}^{\infty} r^n (a_n \cos n\theta + b_n \sin n\theta) \\
&= \sum_{n=1}^{\infty} r^n \left(\left\{ \frac{1}{nR^{n-1}\pi} \int_0^{2\pi} g(\phi)\cos n\phi\, d\phi \right\} \cos n\theta + \left\{ \frac{1}{nR^{n-1}\pi} \int_0^{2\pi} g(\phi)\sin n\phi\, d\phi \right\} \sin n\theta \right) \\
&= \frac{1}{\pi} \sum_{n=1}^{\infty} \left(\frac{R^{1-n}}{n} \right) r^n \left\{ \int_0^{2\pi} g(\phi)\cos n\phi \cos n\theta\, d\phi + \int_0^{2\pi} g(\phi)\sin n\phi \sin n\theta\, d\phi \right\} \\
&= \frac{1}{\pi} \sum_{n=1}^{\infty} \left(\frac{R^{1-n}}{n} \right) r^n \left\{ \int_0^{2\pi} g(\phi)\cos[n(\theta-\phi)]\, d\phi \right\} \\
&= \frac{1}{\pi} \int_0^{2\pi} g(\phi) \left\{ \sum_{n=1}^{\infty} \left(\frac{R^{1-n}}{n} \right) r^n \cos[n(\theta-\phi)] \right\} d\phi
\end{aligned}$$

Green's Functions

A Green's function $G(z)$ in a region Ω is a harmonic function at all points $z \in \Omega$ except at the point $z = z_0$, which is a logarithmic pole. Therefore $G(z) + \ln|z - z_0|$ is harmonic for all $z \in \Omega$. In addition, $G(z) = 0$ on the boundary $\partial\Omega$ of Ω.

The Green's function satisfies

$$\frac{\partial^2 G}{\partial x^2} + \frac{\partial^2 G}{\partial y^2} = \delta(x - x_0, y - y_0), G(x,y,x_0,y_0) = 0 \qquad \text{if } (x,y) \in \partial\Omega \tag{12.6}$$

where $\delta(z - z_0)$ is the Dirac delta function. We consider three fundamental cases. The Green's function in the upper half plane with singularity at $z = z_0$ is

$$G(z, z_0) = \frac{1}{2\pi} \ln \left| \frac{z - z_0}{z - \bar{z}_0} \right| \tag{12.7}$$

Using Eq. (12.7), we can obtain the Green's function for the quarter plane $0 < x < \infty$, $0 < y < \infty$ by using the map $w = z^2$. After some algebra you can show that

$$\begin{aligned}
G(z, z_0) &= \frac{1}{2\pi} \ln \left| \frac{z^2 - z_0{}^2}{z^2 - \bar{z}_0{}^2} \right| \\
&= \frac{1}{4\pi} \ln \frac{\{(x - x_0)^2 + (y - y_0^2)\}\{(x + x_0)^2 + (y + y_0^2)\}}{\{(x - x_0)^2 + (y + y_0^2)\}\{(x + x_0)^2 + (y - y_0^2)\}}
\end{aligned}$$

Finally, the Green's function for the unit disk (with singularity z_0 inside the unit disk) is given by

$$G(z, z_0) = \frac{1}{2\pi} \ln \left| \frac{z - z_0}{1 - \bar{z}_0 z} \right| \tag{12.8}$$

EXAMPLE 12.6

Consider the strip $0 < y < \pi, -\infty < x < \infty$ and find the Green's function for Laplace's equation with Dirichlet boundary conditions.

SOLUTION

The strip is shown in Fig. 12.4.

The strip can be mapped to the upper half plane using $w = e^z$. We can write down the Green's function for the strip immediately using Eq. (12.7) together with $w = e^z$:

$$G(z, z_0) = \frac{1}{2\pi} \ln \left| \frac{e^z - e^{z_0}}{e^z - e^{\bar{z}_0}} \right|$$

EXAMPLE 12.7

Find the Green's function for the half disk $0 < r < 1, 0 < \theta < \pi$.

SOLUTION

The problem can be done by using conformal mapping twice. The first map we can apply takes the half disk to the quarter plane:

$$w = i\frac{1 - z}{1 + z} \tag{12.9}$$

This is illustrated in Fig. 12.5.

A second mapping can be applied to take the quarter plane to the half plane. This is $W = w^2$. This is shown in Fig. 12.6.

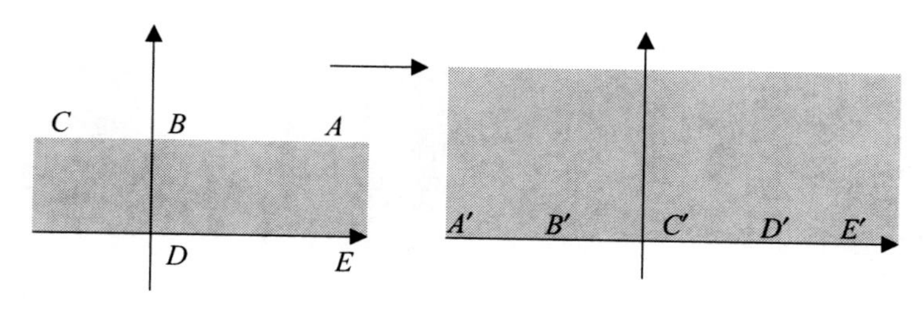

Figure 12.4 A strip of height $B = \pi$ is to be mapped to the upper half plane in Example 12.6.

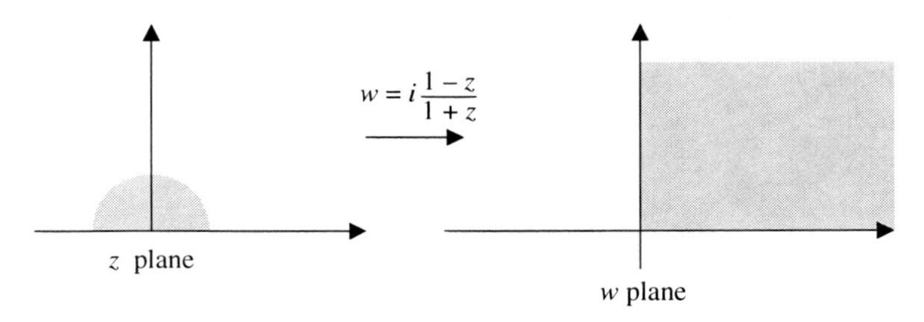

Figure 12.5 The mapping $w = i(1-z)/(1+z)$ takes the half disk to the quarter plane.

Now we have the half plane and can apply what we already know. Using Eq. 12.7, we have

$$G(W, W_0) = \frac{1}{2\pi} \ln \left| \frac{W - W_0}{W - \overline{W}_0} \right|$$

Reversing course, we get the Green's function for the quarter plane:

$$G(w, w_0) = \frac{1}{2\pi} \ln \left| \frac{w^2 - w_0^2}{w^2 - \overline{w}_0^2} \right|$$

Finally, to get the Green's function for the half-disk, we utilize Eq. (12.9), $w = i(1-z)/(1+z)$. This expression is just substituted for w in $G(w, w_0)$. Some tedious algebra gives the final answer:

$$G(z, z_0) = \frac{1}{2\pi} \ln \left| \frac{\dfrac{z - z_0}{\left(1 + z_0^2\right)} \left(1 + \overline{z}_0^2\right)}{z + z\overline{z}_0^2 - z_0 - \overline{z}_0 z^2} \right|$$

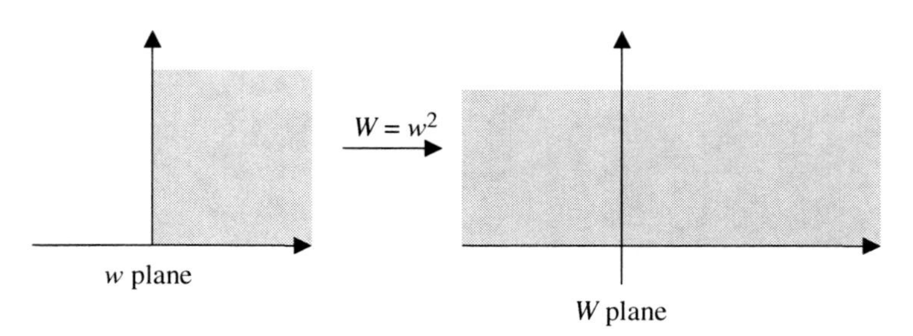

Figure 12.6 The transformation $W = w^2$ maps the quarter plane to the half plane.

Summary

In this chapter, we introduced the application of conformal mapping to the solution of Laplace's equation. It can be applied by using conformal mapping to transform any region into a region with a known solution such as the upper half plane. The idea of Green's functions was also introduced, and the use of conformal maps to write down Green's functions for Laplace's equation in different regions was described.

Quiz

1. Find the harmonic conjugate of the Poisson kernel $P(r,\theta) = \frac{1}{2\pi}\frac{1-r^2}{1-2r\cos\theta + r^2}$.
2. Find the harmonic conjugate of $u(x,y) = \frac{1}{2}\ln(x^2 + y^2)$.
3. Suppose that $\phi(u,v) = v$ in the horizontal strip $-\pi/2 < v < \pi/2$. Is the function harmonic? Find a map that maps the right half plane $x > 0$ in the z plane onto this strip and find a function that is harmonic on the right half plane.
4. Solve $\Phi_{xx} + \Phi_{yy} = 0, y > 0$ if $\Phi(x,0) = \begin{cases} A & x < -1 \\ B & -1 < x < 1 \\ C & x > 1 \end{cases}$
5. Find the Green's function for a triangular wedge in the quarter plane with angle α (Hint: The transformation z^k expands angles, choose a transformation to take $\alpha \to \pi$ to cover the entire upper half plane).

Final Exam

1. Using the definition of the derivative of a complex function in terms of limits, find $f'(z)$ when $f(z) = z^3$.
2. If $f(z) = z$ find $\dfrac{\Delta w}{\Delta z}$ and determine if the function is differentiable.
3. Find the derivative of $f(z) = 3z^2 - 2z$.
4. Find the derivative of $f(z) = \dfrac{z^3}{1+z^2}$.
5. Find the derivative of $f(z) = (2z^2 + 2i)^5$.
6. Is $f = 2x + ixy^2$ analytic?
7. Is the function $f = (x+iy)^3 - (x^2 - i2xy - y^2)(x+iy)$ analytic?
8. Find $f'(z)$ when $f = \sqrt[3]{r}\, e^{i\theta/2}$.
9. In what domain is $f = \cosh x \cos y + i \sinh x \sin y$ analytic?
10. Let $f = u + iv$ be analytic. Show that v is harmonic.

11. Consider the gamma function. Starting with the standard definition, $\Gamma(z) = \int_0^\infty t^{z-1} e^{-t} dt$, use $t = \ln\left(\frac{1}{u}\right)$ to reexpress the gamma function in terms of the natural logarithm.
12. Find the harmonic conjugate of $u = \dfrac{x^2 - y^2}{(x^2 - y^2)^2 + 4x^2y^2}$.

In questions 13 to 25, use integral theorems from complex variables to evaluate the following integrals.

13. $\displaystyle\int_0^\pi \frac{dx}{2+\sin^2 x}$
14. $\displaystyle\int_0^\pi \frac{dx}{5+4\cos x}$
15. $\displaystyle\int_{-\infty}^\infty \frac{\cos x - \cos a}{x^2 - a^2} dx$
16. $\displaystyle\int_0^\infty \frac{x^3 dx}{(x^3+2)^2}$
17. $\displaystyle\int_{-\infty}^\infty \operatorname{sech}^2 x \cos x \, dx$
18. $\displaystyle\int_{-\infty}^\infty \frac{e^{i2x}}{1+4x^2} dx$
19. $\displaystyle\frac{1}{2}\int_0^{2\pi} \frac{d\theta}{1+\cos^2\theta}$
20. $\displaystyle\int_0^\infty \frac{x \sin x}{(x^2+3)(x^2+4)} dx$
21. $\displaystyle\int_0^\infty \frac{\sqrt{x}}{(x+3)^2} dx$
22. $\displaystyle\frac{1}{2\pi}\int_{-\infty}^\infty \frac{\sin x}{x(x^2+4)} dx$
23. $\displaystyle\int_0^\infty e^{-x^2} \cos 2x \, dx$
24. $\displaystyle\int_0^\infty \frac{\cos x \, dx}{1+x^2}$
25. $\displaystyle\int_0^\pi \frac{d\theta}{6-3\cos\theta}$

Write the following in the standard form $z = x + iy$.

26. $(1 + 2i)^3$
27. $(1 + i)^3$
28. $\dfrac{6i + z\overline{z}}{3}$
29. What are the real and imaginary parts of $1 + z^2\overline{z}$?
30. What is the modulus of $1 + z^2\overline{z}$?
31. Find the modulus of $z = (2 + i)^2$.
32. Find the modulus of $z = \dfrac{1-2i}{3}$.
33. What is the residue of at $f(z) = \dfrac{1}{z}$ at $z = 0$?
34. What is the residue of $f(z) = \dfrac{1}{z^3}$ at $z = 0$?
35. Find the residue of at $\dfrac{\sin z}{1+z^3}$ $z = -1$.
36. What are the singularities of $f(z) = \dfrac{z^2}{(z^2+1)(z+4)}$?
37. What is the residue of $f(z) = \dfrac{z^2}{(z^2+1)(z+4)}$ for $z = -4$?
38. What is the residue of $f(z) = \dfrac{z^2}{(z^2+1)(z+2)^2}$ when $z = -2$?
39. Find the singularities and their order of $f(z) = \dfrac{z^2}{z(z+4)(z-2)^2}$.

Find the Laplace transform of the following:

40. $e^{-x} \cos x$
41. e^{ix}
42. $\sinh x$
43. $t^2 e^{-1}$

If possible, use the Bromvich inversion integral to find the inverse Laplace transform of the following:

44. $\dfrac{s}{s^2+1}$
45. $\dfrac{1}{s^2-1}$

46. $\dfrac{1}{s^2+2}$

47. $\dfrac{s}{s^2+2}$

48. $\dfrac{1}{s^2+4s-8}$

49. $\dfrac{s}{s(s^2+1)(s-3)}$

50. $\dfrac{4}{s^5}$

Find the series expansions of the following functions about the origin. Identify the principal part if it exists:

51. $\tan z$

52. e^z

53. $\dfrac{e^z}{z^3}$

54. $\cos z$

55. $\dfrac{\sin z}{z^2}$

56. $\dfrac{\sinh z}{z}$

57. $\dfrac{\cosh z}{z^3}$

58. $\ln|z-2|$

59. $\dfrac{1}{z}\ln\left|\dfrac{z+1}{z-1}\right|$

60. $\dfrac{1}{z(z+1)(z-2)^2}$

Calculate the following limits:

61. $\lim_{z\to 2i} z^2$

62. $\lim_{z\to 2} z^2$

63. $\lim_{z\to 1+i} \bar{z}$

64. $\lim_{z \to \pi} e^{iz}$

65. $\lim_{z \to \pi} \frac{\cos z}{z}$

66. $\lim_{z \to 0} \frac{\sin \pi z}{\pi z}$

67. $\lim_{z \to \infty} \frac{z}{z^2 + 6}$

68. $\lim_{z \to 0} \frac{z}{\sin z}$

69. $\lim_{z \to 1} \frac{z}{(z-1)^2}$

70. $\lim_{z \to 2i} \frac{z}{(z-1)^2}$

Write the following as polynomials in z and $\overline{z}$:

71. $x + y^2$
72. x^3
73. $2x - iy$
74. $2x + 6iy$
75. $4y^2$

Compute the following derivatives:

76. $\frac{\partial}{\partial z}(x^2 + y)$

77. $\frac{\partial}{\partial z}(x^2 + y)$

78. $\frac{\partial}{\partial x}(z^3)$

79. $\frac{\partial}{\partial x} z^2$

80. $\frac{\partial}{\partial x} |z|^2$

Find the real and imaginary parts of the following functions:

81. $f(z) = z\overline{z}$
82. $f(z) = z^2$

83. $f(z) = e^z$

84. $f(z) = \dfrac{1}{z}$ (in polar coordinates)

85. $f(z) = z^{1/3}$

For each of the following functions, indicate where the function may not be analytic.

86. $f = \dfrac{1}{1+z^2}$

87. $f = \dfrac{\sin z}{z}$

88. $f = 1 - z^2$

89. $f = 1 - z^2\bar{z}$

90. $f = \dfrac{2z}{(1-z)(z+2)}$

91. Find the harmonic conjugate function of $u = y^3 - 3x^2 y$.

92. Evaluate $\exp\left(\dfrac{1}{2} + i\dfrac{\pi}{4}\right)$.

93. Evaluate $\exp(2 + 3\pi i)$.

94. Evaluate $\log 1$.

95. Calculate $(1 + i)^i$.

96. Find $e^{i(2n+1)}$ where n is an integer.

97. Evaluate $\sin\left(z + \dfrac{\pi}{2}\right)$.

98. Find $\sin iy$.

99. Find $\cos iy$.

100. Find the roots of the equation $\sin z = \cosh 4$.

Quiz Solutions

Chapter 1

1. 1/8
2. $\frac{1}{5}-\frac{7}{5}i$
3. $z+w=5+2i,\ zw=9+7i$
4. $\bar{z}=2-3i, \bar{w}=3+i$
5. $-3\pi/4$
6. $3\cos^2\theta\sin\theta-\sin^3\theta$
7. $\sin x\cosh y+i\cos x\sinh y$
8. $-i\ln(z\pm\sqrt{z^2-1})$
9. $\left(\frac{\sqrt{3}}{2}+i\frac{1}{2}\right),\left(-\frac{\sqrt{3}}{2}+i\frac{1}{2}\right),-i$
10. $8e^{i\pi/2}$

Chapter 2

1. $3i$
2. $2+i$
3. $\bar{z}^2 + 2z\bar{z} - i$
4. $\dfrac{f+\bar{f}}{2} = \dfrac{z^2+\bar{z}^2+z+\bar{z}+6}{2}$
5. $u(x,y) = 1+x^2+y^2, v(x,y) = -y$
6. 4
7. 1
8. $-i$
9. 0
10. No, $f(1)$ does not exist.

Chapter 3

1. nz^{n-1}
2. $\dfrac{\Delta w}{\Delta z} = \bar{z} + \overline{\Delta z} + z\dfrac{\overline{\Delta z}}{\Delta z}$, no.
3. $24z^7 - 12z$
4. $-6\dfrac{2z^2+3z}{z^4}$
5. $-\dfrac{i}{3}$
6. $u_x = e^x \cos y \neq v_y$
7. $u_x = 1, v_y = -1,$ so the Cauchy-Riemann equations are never satisfied, not even at the origin. So it is not differentiable.
8. Yes
9. The Cauchy-Riemann equations are satisfied, so the derivative exists everywhere in the specified domain.
10. Yes, $v(x,y) = 2xy$

Chapter 4

1. Use the same steps applied in Example 4.5.
2. Consider $e^z = -1$ where $e^x = 1, y = \pi(2n+1), n = 0, \pm 1, \pm 2, \ldots$.
3. e^2
4. $1 + \tan^2 z = \sec^2 z$
5. $\dfrac{\tan z + \tan w}{1 - \tan z \tan w}$
6. Yes, they must be multi-valued, because they are defined in terms of the natural log function.

Chapter 5

1. $N = 2\dfrac{|z|}{\varepsilon}$
2. $\cos\left(\dfrac{n\theta}{2}\right)\dfrac{\sin\left(\dfrac{(n+1)\theta}{2}\right)}{\sin(\theta/2)}$
3. π
4. 4
5. Uniformly convergent to 0, $|z| \geq 2$.
6. $\displaystyle\sum_{n=0}^{\infty} \frac{(-1)^n}{3^{2n+2}} z^{4n+1}$
7. Converges absolutely
8. $\displaystyle -\sum_{n=0}^{\infty} \frac{(z-\pi i)^{2n+1}}{(2n+1)!}$
9. $\displaystyle\sum_{n=0}^{\infty} \frac{(r/2)^{2n}}{(n!)^2}$
10. $\displaystyle\sum_{n=-\infty}^{-1} z^n + \sum_{n=0}^{\infty} \frac{(n+1)}{2^{n+2}} z^n$
11. Removable singularity.

Chapter 6

1. $-5/4$
2. $\dfrac{\sqrt{3}+i}{4}$
3. $\left(\dfrac{1+i}{2}\right)(i+e^{\pi/2})$
4. $\dfrac{1+e^{\pi}}{2}$
5. 0
6. 0
7. arctan (x)
8. $\dfrac{2\pi}{5}$
9. $-\pi i$
10. $-\dfrac{8\pi i}{\pi^2+64}$

Chapter 7

1. 0
2. $i\dfrac{\pi}{3}$
3. $\dfrac{1}{9}\dfrac{1}{(z+1)^2}+\dfrac{2}{3}\dfrac{1}{z+1}$
4. Singularities: $0, -5\pi/2$, residues: $0, 2/5\pi$
5. Singularities: 0, π, residues: $1/\pi$, 0
6. $\dfrac{\pi}{3\sqrt{15}}$
7. $\pi/2$
8. π
9. $\dfrac{5\pi}{6}$
10. $\dfrac{\pi}{e}$

Chapter 8

1. $\dfrac{s}{s^2+\omega^2}$
2. $\dfrac{s}{s^2-a^2}$
3. $f(t)=\dfrac{1}{\sqrt{\pi t}}e^{-k^2/4t}$
4. $\cos \omega t$
5. $f(x)=\left(\dfrac{\sin \alpha\pi}{\pi}\right)t^{\alpha-1}(-\alpha)!$

Chapter 9

1. A parabola described by $u=\left(\dfrac{v}{2a}\right)^2-a^2$
2. $x=a$ is mapped to a circle $|w|=e^a$
3. $w=\dfrac{i-z}{i+z}$
4. $z=2\pm\sqrt{2}$
5. $w=\dfrac{i-z}{i+z}$
6. $Tz=\dfrac{z-1}{z+1}$

Chapter 10

1. It maps an infinitely high vertical strip with $v\geq 0$ of width $W=A\pi+B$ to the upper half plane.

Chapter 11

1. 1
2. 6
3. $\frac{64}{21}\sqrt{\frac{2}{\pi}}\left[\Gamma\left(\frac{1}{4}\right)\right]^2$
4. 2π
5. $\frac{\Gamma\left(\frac{1}{3}\right)}{2\sqrt{3}}$
6. $\Gamma(z)\Gamma(-z) = -\frac{\pi}{z\sin\pi z}, \Gamma(-1/2) = -2\sqrt{\pi}$
7. $\frac{\Gamma(\alpha+1)}{\Gamma(\alpha-n+1)}$
8. This gives the harmonic series.
9. 1
10. The function is entire—it is analytic everywhere in the complex plane.

Chapter 12

1. $v(r,\theta) = \frac{r}{\pi}\frac{\sin\theta}{1-2r\cos\theta+r^2}$
2. $v(x,y) = \arctan\left(\frac{y}{x}\right) + v_0$ where v_0 is a constant
3. Yes, $w = \ln z, U(x,y) = \arctan\left(\frac{y}{x}\right)$.
4. $\frac{A-B}{\pi}\tan^{-1}\left(\frac{y}{x+1}\right) + \frac{B-C}{\pi}\tan^{-1}\left(\frac{y}{x-1}\right) + C$
5. $G(z,z_0) = \frac{1}{2\pi}\ln\left|\frac{z^{\pi/\alpha} - z_0^{\pi/\alpha}}{z^{\pi/\alpha} - \overline{z}_0^{\pi/\alpha}}\right|$

Final Exam Solutions

1. Use $(x+y)^3 = x^3 + 3x^2y + 3xy^2 + y^3$ to get $f'(z) = 3z^2$.
2. 1, yes.
3. $6z - 2$
4. $\dfrac{3z^2 + z^4}{(1+z^2)^2}$
5. $20z(2z^2 + 2i)^4$
6. No
7. No, note that $f(z) = z^3 - \overline{z}^2 z$
8. $\dfrac{1}{3(\sqrt[3]{r}\, e^{i\theta/2})^2}$

9. The function is entire.
10. $u_{xy} = v_{yy}, u_{xy} = -v_{xx} \Rightarrow v_{xx} + v_{yy} = 0$
11. $\int_0^1 \left[\ln\left(\frac{1}{u}\right) \right]^{z-1} du$
12. $v = \dfrac{-2xy}{(x^2 - y^2)^2 + 4x^2y^2}$
13. $\dfrac{\pi}{\sqrt{6}}$
14. $\pi/3$
15. $-\dfrac{\pi \sin a}{a}$
16. $\dfrac{\sqrt[3]{2}\,\pi}{9\sqrt{3}}$
17. $\dfrac{2\pi}{\sinh \pi}$
18. $\dfrac{\pi}{2e}$
19. $\dfrac{\pi}{\sqrt{2}}$
20. $\dfrac{\pi}{e^2}(e^{2-\sqrt{3}} - 1)$
21. $\dfrac{\pi}{2\sqrt{3}}$
22. $\dfrac{\sinh(1)}{4e}$
23. $\dfrac{\sqrt{\pi}}{2e}$
24. $\dfrac{\pi}{2e}$
25. $\dfrac{\pi}{3\sqrt{3}}$
26. $-3 + 4i$
27. $-2 + 2i$
28. $\dfrac{x^2 + y^2}{3} + 2i$

29. $\text{Re} = 1 + x^3 + xy^2, \text{Im} = yx^2 + y^3$
30. $\sqrt{(1 + x^3 + xy^2)^2 + (yx^2 + y^3)}$
31. 5
32. $\dfrac{\sqrt{5}}{3}$
33. 1
34. 0
35. $-\dfrac{\sin(1)}{3}$
36. $z = \pm i, z = -4$
37. 16/17
38. –4/25
39. $z = 0, -1$ order 1, $z = 2$ order 2
40. $\dfrac{1+s}{1+(1+s)^2}$
41. $\dfrac{1}{s-i}$
42. $\dfrac{1}{s^2-1}$
43. $\dfrac{2}{(1+s)^3}$
44. $\dfrac{e^{-t}}{2}(1+e^{2t})$
45. $\sinh t$
46. $\dfrac{\sin\sqrt{2}t}{\sqrt{2}}$
47. $\cos\sqrt{2}t$
48. $\dfrac{e^{-2(1+\sqrt{3})t}}{4\sqrt{3}}(e^{4\sqrt{3}t} - 1)$
49. $\dfrac{1}{10}(e^{3t} - \cos t - 3\sin t)$
50. $\dfrac{t^4}{6}$
51. $z + \dfrac{z^3}{3} + \dfrac{2z^5}{15} + \cdots$

52. $\frac{1}{z}+1+\frac{z}{2}+\frac{z^2}{6}+\frac{z^3}{24}+\cdots$, principal part $\frac{1}{z}$

53. $\frac{1}{z^3}+\frac{1}{z^2}+\frac{1}{2z}+\frac{1}{3!}+\frac{z}{4!}+\frac{z^3}{5!}+\cdots$, principal part $\frac{1}{z^3}+\frac{1}{z^2}+\frac{1}{2z}$

54. $1-\frac{z^2}{2!}+\frac{z^4}{4!}-\cdots$

55. $\frac{1}{z}-\frac{z}{3!}+\frac{z^3}{5!}-\frac{z^5}{7!}+\cdots$, principal part $\frac{1}{z}$

56. $1+\frac{z^2}{3!}+\frac{z^4}{5!}+\cdots$

57. $\frac{1}{z^3}+\frac{1}{2z}+\frac{z}{4!}+\frac{z^3}{6!}+\cdots$, principal part $\frac{1}{z^3}+\frac{1}{2z}$

58. $i\pi+\ln 2-\frac{z}{2}+\frac{z^2}{8}-\cdots$

59. $\frac{i\pi}{z}+2+\frac{2z^2}{3}+\frac{2z^4}{5}+\cdots$, principal part $\frac{i\pi}{z}$

60. $\frac{1}{4z}+\frac{3z}{16}-\frac{z^2}{16}+\frac{9z^3}{64}+\cdots$, principal part $\frac{1}{4z}$

61. -4

62. 4

63. $1-i$

64. -1

65. $-1/\pi$

66. 1

67. 0

68. 1

69. ∞

70. $-\frac{2}{25}(4+3i)$

71. $\dfrac{z+\overline{z}}{2}-\dfrac{z^2}{4}+\dfrac{z\overline{z}}{2}-\dfrac{\overline{z}^2}{4}$
72. $\dfrac{1}{8}(z^3+3z^2\overline{z}+3z\overline{z}^2+\overline{z}^3)$
73. $\dfrac{z+3\overline{z}}{2}$
74. $z(1+3i)+\overline{z}(1-3i)$
75. $z^2-2z\overline{z}+\overline{z}^2$
76. $x+\dfrac{1}{2i}$
77. $x-\dfrac{1}{2i}$
78. $3x^2+i6xy-3y^2$
79. $2x+i2y$
80. $2x$
81. $u=x^2+y^2, v=0$
82. $u=x^2-y^2, v=2xy$
83. $u=e^x\cos y, v=e^x\sin y$
84. $u=\dfrac{\cos\theta}{r}, v=\dfrac{\sin\theta}{r}$
85. $u=r^{1/3}\cos\theta/3, v=r^{1/3}\sin\theta/3$
86. Analytic except at $z=\pm i$
87. Analytic except at $z=0$
88. Is entire
89. Not analytic, depends on $\overline{z}$
90. Analytic except at $z=1,-2$
91. $v=x^3-3xy^2$
92. $\sqrt{\dfrac{e}{2}}(1+i)$
93. $-e^2$
94. $2n\pi i \quad n=0,\pm1,\pm2,\dots$
95. $\exp\left(-\dfrac{\pi}{4}+2n\pi+\dfrac{i}{2}\ln 2\right)$

96. $(-1)^{1/\pi}$
97. $\cos z$
98. $i \sinh y$
99. $\cosh y$
100. $\left(\frac{\pi}{2} + 2n\pi\right) \pm 4i, \quad n = 0, \pm 1, \pm 2, \ldots$

Bibliography

Mark J. Ablowitz and Athanassios S. Fokas, *Complex Variables: Introduction and Applications*, 2d ed., Cambridge University Press, Cambridge, U.K. (2003).

James W. Brown and Ruel V. Churchill, *Complex Variables and Applications*, 7th ed., McGraw-Hill, New York (2004).

Robert E. Greene and Steven G. Krantz, *Function Theory of One Complex Variable*, 2d ed., Graduate Studies in Mathematics Vol. 40, American Mathematical Society, Providence, Rhode Island (2002).

Liang-Shin Hahn and Bernard Epstein, *Classical Complex Analysis*, Jones and Bartlett, Sudbury, Massachusetts (1996).

Norman Levinson and Raymond M. Redheffer, *Complex Variables*, McGraw-Hill, New York (1970).

Murray R. Spiegel, *Schaum's Outlines: Complex Variables with an Introduction to Conformal Mapping and Its Applications*, McGraw-Hill, New York (1999).

INDEX

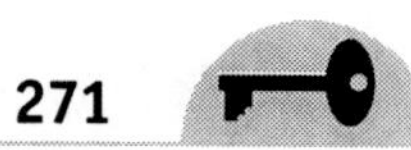

G

H

I

J

K

L

CPSIA information can be obtained at www.ICGtesting.com
Printed in the USA
BVOW06s1818101213

338628BV00005BA/25/P